SYNCHRONICITY

science, myth, and the trickster

ALLAN COMBS & MARK HOLLAND

MARLOWE & COMPANY
NEW YORK

Second edition, 1996
Published in the United States by
Marlowe & Company
632 Broadway, Seventh Fl.
NY, NY 10012

Designed by Kathy Kikkert

Manufactured in the United States of America

Library of Congress Cataloging-in-Publication Data
Combs, Allan, 1942–
    Synchronicity : science, myth, and the trickster /
Allan Combs, Mark Holland.—2nd edition
    Includes bibliographical references.
    ISBN 1-56924-845-1
    1. Coincidence.   2. Science—Philosophy.
3. Mythology.
I. Holland, Mark.   II. Title.
BD595.C65   1996
123—dc20   89-25590
                                                   CIP

There exists a type of phenomenon, even more mysterious than telepathy or precognition, which has puzzled man since the dawn of mythology: the seemingly accidental meeting of two unrelated causal chains in a coincidental event which appears both highly improbable and highly significant.

ARTHUR KOESTLER

Hermes, for to you beyond all other gods it is dearest to be man's companion . . .

The Iliad

CONTENTS

PREFACE

In the fall of 1899, Winston Churchill, then only twenty, was acting as special correspondent to the *Morning Post*. His job was to report on the Boer War in South Africa. On November 14, the armored train he was traveling on was attacked by Boers. Some of the cars were derailed, and the ensuing fight continued vigorously for more than two hours. Though Churchill played a leading role in organizing a successful retreat, he and a number of British soldiers were captured. In the meantime he had somehow escaped injury despite repeatedly exposing himself at close range to enemy marksmen.

Churchill was placed in confinement in the city of Pretoria. Soon, he and several other prisoners set in motion an escape plan. The others, however, backed out in the final moments, leaving only Churchill to continue. After being lowered over the wall of the compound in the middle of the night, he walked straight across the garden, through the outer gate, by the sentry, and out of town. Of the two thousand imprisoned in Pretoria, he was the only one to escape. After several days on his own, however, his situation grew desperate. Exhausted, hungry, and with no clear idea of his location—Pretoria was, in fact, three hundred miles from the British border—he decided one night to approach some lights in the distance and take his chances. It was a mining camp. Knocking at a door, he found himself in the company of John Howard, the only English sympathizer within twenty miles. Howard was able to smuggle Churchill out of enemy territory, and he arrived home a hero.[1]

Some forty years later, Churchill was to become the great adversary of Adolf Hitler, who like himself had also experienced war firsthand. Hitler served in the German infantry during the First World War. As a courier it often was his job to carry messages along battle lines during vigorous fighting—a dangerous assignment, but one that he seemed to thrive on. Indeed, he seemed to live a charmed life. Once he walked out of his commander's headquarters just before it was hit by an English artillery shell which killed three persons and seriously wounded the commander. Time and time again Hitler seemed to come within a hair's breadth of death and escaped unharmed. This ability was to stay with him throughout his life. He later wrote to a reporter about one episode from his combat experience: "Four times we advanced and had to go back; from my whole batch only one remains, beside me; finally he also falls. A shot tears off my right coat sleeve, but like a miracle I remain safe and alive."[2]

Both Churchill and Hitler seemed to live under the aegis of fate, as if protected so that they might later play their roles in history. The vehicle for fate was the operation of chance or coincidence. Churchill, by what appears to be the operation of fortuitous coincidence, placed himself in the hands of an English sympathizer. Bullets and artillery shells seemed to land everywhere except on Adolf Hitler. Perhaps, however, their lives did not differ in a fundamental way from our own. We as individuals also experience occasional but dramatic interventions of coincidence. Such interventions produce an immediate and palpable sense that something more than blind chance is working behind the scenes of life's dramas.

Often, however, it is not from the dramatic visitations of chance, but from the smallest of coincidences that we feel some agency to be operating behind the scenes. This agency asserts itself in the most random of events, stamping them with an uncanny intelligence that can only be called purposeful. For instance, you wake up one morning thinking of someone you have not seen in years, only to meet him in an elevator later the same day. In the meantime, a mutual acquaintance has inquired about him. Or you set off for the library with the intention of looking up books on greenhouses as well as on cheeses. At a drugstore on the way you find a magazine on the newsstand with feature articles on both of these topics.

While we may conclude that such coincidences are produced by nothing more than the endless shufflings and reshufflings of random everyday events, the frequency with which they occur belies such an interpretation. Occasionally you experience a sequence of coincidences so dramatic as to make such an explanation seem nothing short of silly. For example, the story is told of a certain M. Deschamps who, as a boy in Orleans, France, was presented with a piece of plum pudding by a guest of the family, M. de Fortgibu. Years later Deschamps, now a young man, ordered plum pudding in a Paris restaurant, only to find that the last piece had just been taken. The waiter discretely indicated the direction of the guilty patron who, it turned out, was none other than de Fortgibu. Many years later, at a dinner party where Deschamps was again offered plum pudding, he took the opportunity to recount the above events concerning de Fortgibu. Finishing his tale, and still eating his plum pudding, he remarked that all that was missing was de Fortgibu. Soon the door burst open and in came de Fortgibu himself, now a disoriented old man who had gotten the wrong address, and so had entered by mistake![3]

The most common meaningful coincidences are those seemingly random but apparently purposeful events which speak to us directly in terms of personal meaning. Psychiatrist Carl Jung, for instance, reported the arrival of a beetle at the window of his office as a client was describing a dream involving just such an insect. Her dream concerned the golden Egyptian scarab beetle:

> While she was telling me this dream I sat with my back to the closed window. Suddenly I heard a noise behind me, like a gentle tapping. I turned around and saw a flying insect knocking against the window-pane from outside. I opened the window and caught the creature in the air as it flew in. It was the nearest analogy to a golden scarab that one finds in our latitudes.[4]

In Egyptian mythology this beetle is a symbol of rebirth, and its appearance in the dream marked a critical development in this woman's therapy. Jung notes that up to that time she had been extremely difficult to treat, clinging to a rigid notion of reality with such tenacity that two previous psychiatrists had been unable to dislodge it. "Evidently something quite irrational was needed which was beyond my powers to produce."[5] The dream did part of this work, but when the actual insect came flying through the window, the whole rigid structure of her perception of reality began to change.

Jung used the term *synchronicity* for such meaningful coincidences. His investigation of coincidences that occurred in his own life and in the lives of others led Jung to conclude that they are related to unconscious psychological processes. He did not, however, stop with the psychological side of synchronicity. He worked with his friend Wolfgang Pauli, the great quantum physicist who explicated the Pauli exclusion principle, to develop the notion that the laws of physics themselves should be rewritten to include *acausal* as well as causal accounts of the world of physical events.

This book is an exploration of synchronicity. If we, like Jung, take synchronicity seriously and begin to examine its implications, we will, like him, be led to a fundamental reexamination of human nature, the nature of the physical universe, and the relationship between the two. Coincidences such as that of the golden scarab

beetle bring the mental world of human meaning into direct con-
junction with the world of physical reality, ruled by the laws of biology
and physics. To take a step in the direction of understanding synchronicity, we must revise our traditional views of mind, as well as
our understanding of Nature herself.

Allan Combs
Mark Holland

FOREWORD
TO THE SECOND EDITION

There are two things I have come to believe implicitly about the world we live in. One is that nothing occuring in it is independent of any other thing; the other is that nothing that occurs is entirely random and prey to chance. These two beliefs are part of the same insight: if all occurrences are linked in some way with all others, everything acts in some way on everything else. Nothing happens in a purely random way. This, of course, does not mean that everything that happens is fully determined by all that happens elsewhere or happened before. It does mean, however, that there is no such thing as pure coincidence. When something happens, it happens in some, though possibly extremely subtle, relation to other things that happen, or have happened, within that region of space and time.

Evidently degrees of relevance must be distinguished. Though in this universe all things are connected, they are not all connected to the same degree: *like* is more connected to *like* than to *unlike*.

This may sound like a metaphysical proposition, but it is proven correct already in the "hardest" of all sciences, quantum physics. A photon or an electron turns out to be linked to another photon or electron over finite distances, provided the two particles originated in the same quantum state. The famous EPR, Einstein-Podolski-Rosen, thought experiment, with its more recent empirical test by Alain Aspect, has demonstrated that particles emitted as identical twins remain interlinked even when separated from each other over considerable distances. The wave-function of one particle collapses when a measurement is made on the other. All this is recounted by Allan Combs and Mark Holland in the chapter on physics and synchronicity. That identical twins of the human kind would remain interlinked is likewise established. Thousands of cases have shown that one twin can often feel the pain of the other, and can be aware of a trauma befalling him or her.

Twins, whether electronic or human, are just two kinds of *like* that are linked with *like* in the known world; there are many others. Perhaps the least controversial findings relevant to linkages among humans are those that show that in the state of deep meditation the EEG waves of people who are emotionally close to each other become progressively synchronized even when there is no sensory contact between them.

These remarks introduce the topic of the book now in your hand: *Synchronicity* is a phenomenon of connectivity in human experience, one of the most notable findings of Jungian psychology. It has been plagued, however, by the skeptics' objection that synchronistic events are "nothing but" mere (if often curious) coincidences. Now, nothing-butness has been a bastion of the reductionist's view of reality, where organisms are said to be "nothing but" atoms and molecules and societies and cultures "nothing but" human individuals with particular mind-sets. Fortunately, reductionism has been proven wrong in the new physics, where even the properties of ordinary atoms were shown to be irreducible merely to the properties of their parts. Indeed, atoms have emergent properties and emergent interactions, both among their components and between their components and their surroundings. (The Pauli *exclusion principle*, for example, does not apply to single electrons, nor to sets of electrons unless they are associated with atomic nuclei; and chemical valence is not a property of electrons and nucleons, only of the whole atom that is constituted of them.) Irreducibility applies with equal force to the biological realm. (For instance, there is no trace of the coordination and control of the biochemical reactions that characterize a whole organism within the individual cells and molecules that make it up.) This concept applies with still greater force to the whole human being, where the processes that give rise to consciousness are a collective property of the complex neural nets that form the brain, and cannot be traced to the properties of the individual neurons within those nets.

If it is already evident that nothing-butness is an obsolete assumption with respect to organized wholes in nature, it is far less evident that nothing-butness is equally obsolete in regard to synchronistic events. Yet the basic assumption is similar in the two cases. In both, the principal, though usually unstated, assumption is that what we cannot perceive with our eyes or register with our instruments is not only questionable; it does not exist. It is forgotten that our eyes, even when aided with physical measuring devices, respond to but a small portion of the electromagnetic spectrum, and that what we know of the other universal fields and forces, gravitational and nuclear, is for the most part highly indirect. Indeed, by far the greater portion of what science regards as physical reality is neither directly nor instrumentally observable. It is, however, connected with observable events by rigorously constructed and repeatedly tested rules. For the reality postulate to hold, the connections between theoretically postulated unobservable entities and the entities that are at least partially available for observation must be internally consistent and coherent with the main body of current assumptions about a given region or aspect of the empirical world. We can apply this precept to synchronicity.

It appears that the processes which give rise to synchronistic phenomena are unobservable. But are they real?

Surely, not all surprising coincidences that happen to most of us have a background in reality. But some may have. Realistic synchronicities presuppose an actual connection between otherwise disjoined occurrences. If it could be shown that some, even if not all, events that appear as synchronicities have a realworld interconnection, our entire attitude toward synchronicity would change. Instead of remaining a kind of psychological curiosity, the phenomenon would move into the realm of science. There are new developments in our understanding of the empirical world that augur such a shift. But before reviewing them, we should look at the psychological insights themselves.

Carl Jung, as Combs and Holland show in their fascinating study, regarded meaningful coincidences as having more than merely a subjective meaning for the person who experiences them. He suspected, and did not hesitate to say, that synchronicities have a basis in the objective world to which subjective experience refers. The term he used to describe this aspect of reality is *unus mundi*. This is the unitary ground that brings together domains of experience as distinct as quantum physics and analytic psychology. Jung wrote,

> The common background of microphysics and depth-psychology is as much physical as psychic and therefore neither, but rather a third thing, a neutral nature which can at most be grasped in hints since in essence it is transcendental ... The transcendental psychophysical background corresponds to a "potential world" in so far as all those conditions which determine the form of empirical phenomena are inherent in it.[1]

Our connection with this neutral ground of reality, which Jung also called *psychoid* since it is both psychical and physical, occurs in and through the unconscious mind. It is in the unconscious that the elements of synchronicity surface; it is there that the archetypes that pervade both the individual and the collective unconscious appear.

Jung was rather definite on the underlying nature of synchronicities. They concern the appearance of archetypes, he said, mediated through the collective unconscious. The archetypes that appear are not casually connected with one another: their connections are transcendent, that is, beyond space and time. But archetypal phenomena themselves are real. When we perceive them in the guise of a synchronistic event, we are not exercising some paranormal power of perception; we are reporting on a real occurrence that occurs within our psyche. "I emphasize the reality of the event," Jung wrote in a hitherto untranslated letter written in February, 1966, "but not of its perception. This conception is consistent with the hypothesis of acausal connections, that is, with a nonspatial and nontemporal determination of being."[2] Here *being* (*Sein*) is used to indicate the collective aspect of consciousness: the psyche. Elsewhere Jung noted,

> The psyche can be regarded as an unextended intensity, not a body moving in time.... Our brain might be the place of transformation, where the relatively infinite tensions or intensities of the psyche are tuned down to perceptible frequencies and extensions. But, in itself, the psyche would have no dimension in space and time at all.[3]

Psyche as the unextended collective unconscious is a transcendent reality, beyond space and time, and hence beyond the causality that applies to events in space and time. This is the reality of the *unus mundi*, neutral as regards physical and psychical reality and standing behind both of them equally. Developments in contemporary science, especially in quantum field physics and the biophysics of the organism, deepen, though in some ways also modify these insights.

One example of the most relevant findings concerns the effect of electromagnetic (and more complex) quantum fields on living organisms. Evidently the sensitivity of the living state extends considerably below the thermal threshold of chemical reactions, and electromagnetic fields of extremely low frequency can produce effects, some fields remaining active even at intensities near zero.[4] Such fields may actually act on organisms not only by virtue of their strength, but in terms of the richness of information they contain.[5]

Since the human organism is constantly embedded in an ambient electromagnetic field with complex quantum components, and since this field carries a wide variety of patterns with specific informational significance, it seems possible that additional channels of information are available to the brain. Indeed, human brains are not isolated systems enclosed in craniums and capable of communication only through signals transduced by the five exteroceptive senses. A wide variety of information is likely to be reaching the brain at all times, even if conscious awareness filters out the greater part of it.

Ambient electromagnetic fields encode a rich variety of processes that exist in the world beyond the brain. These encoding fields can be likened to a sea in which every pebble creates a ripple that propagates throughout the surface, interacting with other waves and ripples and producing complex interference patterns. Such highly structured fields resemble super-holograms, where interfering wavefronts code not only the visual outlines of objects, but all aspects of their motion in space and time. In principle, it should be possible to decode every element of information carried by such fields, though doing so would exceed the information-processing capacity of our brains, which in practice are limited to accessing certain kinds of patterns: those consistent with the patterns created by their own neural functions. The two-way translation rules discovered by Fourier and refined by Gabor apply here: the reverse transforms (from fields to brain) are the inverse of the forward transforms (from brain to fields). This means that what we read out from the fields is limited mainly to what we (and others like us) read in.

The patterns carried by electromagnetic and quantum fields in the biosphere constitute a subtle information field. I have called it the *psi-field*, through which we communicate with others of our kind, whether we are consciously aware of it or not.[6] These fields, explained in greater detail in Appendix III, act like Jung's collective unconscious, with the difference that they convey to us in an accessible form not only basic archetypes, but well-nigh all things that are products of human brains. The information encoding capacity of the fields is staggering. This was not known in Jung's time. Had he known that a holographic medium the size of a cube of sugar, for example, can encode all the information contained in the U.S. Library of Congress, he would most likely have enlarged his view of the repertory of information in the collective unconscious.

When a notion or image surfaces in our consciousness that seems *acausally* connected with other notions or events, it is likely to be conveyed through the information-rich electromagnetic and quantum fields. With the instruments that are now becoming available for testing field effects, it should be possible to examine when our brain responds to field-coded information originating in brains with similar neural processes across finite distances. And, if the fields store information without short-term attenuation, it should also be possible to see when our brain responds to information that originated in isomorphic brains over finite times. The connections ambient electro magnetic and quantum fields create between events are real, but neither acausal nor transcendent. They are causal in the same sense in which messages in a conversation with a cellular telephone are causal: they mutually affect sender and receiver even though they are transmitted through the spectral domain of the electromagnetic field. And the connections are also immanent inasmuch as the fields that carry them are part of the physical world.

If these hypotheses, which I have outlined in detail elsewhere, correspond to reality, some, even if not all, symbolically or subjectively meaningful coincidences would have a basis in physical reality.[7] In the present book Combs and Holland give a number of intriguing examples, and my Foreword would not be complete unless it added one more case to their collection.

A few weeks before I wrote this note I woke up in the middle of the night with the feeling that I should be hearing from the Malaysian production editor of *World Futures*, the quarterly journal I am editing. After a few minutes I realized, however, that Niki Chong, the editor in question, was not particularly overdue with information about the line-up for the next issue, and thus there was really not much point in worrying about it. The worrying thought did not go away, however, and after a few minutes I put on the light and made a note, "fax Niki Chong about WF line-up." I glanced at the watch; it was 3:35 a.m. Next morning I would have forgotten all about the matter except for the note on the night table. Though I did not see much point in faxing Niki, I picked up the note and went to my office. There was an incoming fax in the machine. It was from Niki Chong, giving the line-up. I looked at the time it was sent from Malaysia: 9:30 a.m. That was exactly 3:30 a.m. in my place in Italy.

Mere coincidence? Perhaps. Then again. . . .

Ervin Laszlo
June, 1995
Pisa, Italy

REFERENCES

1. Jung, C.G. (1970). *Mysterium conjiunctionis: An inquiry into the separation and synthesis of psychic opposites in alchemy.* CW, Vol. XIV, 2nd ed., Princeton, N.J.: Princeton University Press.
2. Jung, C.G. (1961). *Ein Brief zur Frage der Synchronizitit.* Zeitschrift fur Parapsychologie und Grenzgebiete der Psychologie. No.1.
3. von Franz, Marie-Louise. (1992). *Psyche and matter.* Boston & London: Shambhala, p.161.

4. Del Giudice, E.G. , Doglia, S., Milani, M., & Vitiello, G. (1986). In F. Guttmann and H. Keyzer, eds., *Modern Bioelectro-chemistry*. New York: Plenum.
5. Persinger, M.A. (1995). On the possibility of directly accessing every human brain by electromagnetic induction of fundamental algorithms. *Perceptual and Motor Skills*, 80, 791-799.
6. Laszlo, E. (1993). *The Creative Cosmos*. Edinburgh: Floris Books.
7. Laszlo, E. (1995). *The Interconnected Universe*. London and Singapore: World Scientific.

INTRODUCTION:
science, myth, and nature

Pythagoras, it was said, could read
the meaning of ripples of water stirred by the wind. He evidently
believed that seemingly random events in nature form a common
fabric with events in human lives. This may seem odd to us, but
such a belief was not unusual in the ancient world. The Chinese,
for example, read answers to questions scratched onto smooth pieces
of bone or tortoise shell by examining the random patterns of cracks
produced when these objects were placed in a fire. In fact, the notion
that the cosmos is formed of a single common fabric which includes
both the worlds of nature and of humankind, a fabric in which each
event, however insignificant, is connected to each other event, in
varying forms remained with us up to the beginnings of modern
science.

In classical times, for instance, Hippocrates wrote, "There is one
common flow, one common breathing, all things are in sympathy .
. . the great principle extends to the extremist part, and from the

extremist part it returns to the great principle, to the one nature, being and not-being."[6] In the Middle Ages, people believed in a creation in which God played a role in all things, no matter how small. All events were touched and harmonized by this divine purpose and consent. In the late medieval world the various levels of the cosmos, the earth below, the celestial spheres above, and the human soul, were thought to be bound together by mutual *harmonies* or *sympathies*. Events in one were reflected in the others. For the inhabitant of the ancient and medieval worlds, the cosmos possessed the properties of a womb: It enclosed one entirely, supporting and carrying one forward through life in a fashion which, if not always comfortable, was at least meaningful. In the seventeenth century this state of affairs was shattered utterly. The force that shattered it was the relentless hammer of mechanistic science.

Mechanistic science was founded on the belief that the universe is composed of small, unyieldingly solid objects—*atoms*—floating and interacting in an absolute void. Democritus had held a similar view centuries before, declaring that "nothing exists but atoms and void." This notion was not taken seriously by large numbers of people, however, until the philosophers and scientists of the seventeenth century—Francis Bacon, Thomas Hobbes, Johannes Kepler, Galileo Galilei, René Descartes, Isaac Newton, and others—adopted it as the basis of the new and rapidly growing physical sciences of the day. Their century, like our own, was one of discontent with traditional beliefs. Intellectuals, especially those committed to the new sciences, were desperately seeking escape from the straitjacket of religious dogmas handed down through the church from the Middle Ages.

The new mechanistic worldview powerfully canceled out the earlier, more comforting view that saw meaningful connections between apparently discrepant events. It became impossible that coincidences which share no apparent causative agency could form meaningful relationships with each other. Instead, so-called synchronistic events could share only the relationship of cards in a deck which, after being thoroughly shuffled, happen to fall beside each other or to appear in the same poker hand. All this change—from a spiritual to a mechanistic worldview—meant a change in the fundamental myth with which human beings explained their universe.

SCIENTIFIC MYTHOLOGIES

Widely held belief systems constitute *mythologies*, regardless of their scientific or religious origins. According to the historian and astute student of mythology, William Irwin Thompson,[7] myths provide answers to the three questions: What are we? Where do we come from? Where are we going? Along with philosopher Ian Barbour,[8] we would also add the question: What is the real nature of the world; that is, how and of what is it formed?

In late medieval Europe, questions such as these were answered by a powerful mythological system. The grand mythic architecture of creation, handed down from Aristotle and rewritten by Christian philosophers of the thirteenth century, involved a series of mutually enclosed spheres beginning with the earthly world below and ending with the divine empyrean sphere at the top. This vision is described by Dante in the *Divine Comedy*. Networks of resonance, or *sympathies*, formed interconnections throughout this system so that no event occurred in isolation. In this cosmos, meaningful coincidences were natural. For instance, someone dreams of a distant disaster before actually receiving news of it. Or three friends living in separate parts of a large city find themselves drawn to the same restaurant at the same time. A flock of blackbirds lands on a house where someone is dying; howling dogs are heard in the distance. Such coincidences contain common resonances that run through nature and the human soul, drawing both outer and inner reality into shared patterns of meaning. The view of the world provided by religion allowed for the meaningful connection we today call synchronicity. It is as if people in medieval times told themselves stories about the universe which included synchronistic events in the plot structure.

Indeed, myths are stories—Aristotle's use of the word can be translated as "plot"—which tells us about the nature of the realities in which we live. Such stories are grounded in the unconscious, where they act from behind the scenes of perception to structure our understanding of the world. As Alfred North Whitehead, the great English mathematician and philosopher, wrote:

> In every age the Common interpretation of the world of things is controlled by some scheme of unchallenged and unsuspected presupposition: and the mind of any individual, however little he may think himself to be in sympathy with his contemporaries, is not an insulated compartment, but more like one continuous medium—the circumambient atmosphere of his place and time.[9]

Such presuppositions are the stuff of myth. They are the lines with which the great narratives of reality are written. To the individuals that live them, however, they are invisible, as water is to fish. They live in it and it supports them, but they take it entirely for granted. Needless to say, scientists are no more exempt from this state of affairs than is anyone.

Mechanistic seventeenth-century science brought with it new myths, which not only transformed the official science of the day but also changed the maps of the world that each person carried unconsciously within. This science viewed the cosmos as a great vacuum filled with solid atoms that interact only by direct impact. The result, while highly successful for the emerging sciences of physics, astronomy, chemistry, and medicine was devastating to a sense of individual belonging and connection to the cosmos as a whole. Eventually, society and even human nature itself were restructured along the same lines: people came to be seen as separate and distinct objects, connected to each other and to the natural world only where direct contact was possible. The older notions of sympathies were reduced to the status of mere superstitions.

As time passed and human nature was seen increasingly through the eyes of the new science, people took on a further characteristic of atoms. The word "atom" means literally indivisible—that which cannot be broken down or divided. That is to say, there is no access to its interior, if indeed it has an interior at all. Likewise, people came to be viewed increasingly as having no interior and were reduced by mechanistic science to objects driven entirely by external forces. Inside there was nothing. As it emerged in the psychology of the early twentieth century, behaviorism viewed human nature in exactly this way.

Now, the experience of a living and vital interior is the essence of human subjectivity. It is this subjectivity as it mirrors the world without in its own feeling that is the very essence of the human soul. As Novalis wrote in 1798, "The seat of the soul is where the inner world and the outer world meet. Where they overlap, it is in every point of the overlap."[10] This soul is the source of those characteristics we regard as most uniquely human: creativity, intuition, and conscience. It was this soul, this melding of the outer with the inner, to which those in medieval times turned for wisdom, and it was this soul that responded in sympathy to events beyond its immediate surroundings. The loss of the interior dimension of human nature reduced us to two-dimensional beings cast about by events on the physical surface of reality. It terminated our intimate bond with the rest of the cosmos.

Thus science discouraged us from looking into ourselves for the meaning of things. It especially discouraged us from looking below the emotional surface of our own perceptions for felt meanings, the very stuff that formed much of the wisdom of previous ages. Galileo asserted that science should look and see what is actually there. Yet, as physicist Jeremy Haywood points out, "For Galileo only things which could be measured were to be counted as valid grounds for argument, and the language for describing them was to be the pure language of mathematics, free from human wish and whim." Haywood notes that this attitude, built into the foundations of mechanistic science, disclaims "the human capacity to know things directly as they are—fundamentally [it proclaims] the inadequacy of man to grasp complete truth."[11]

In the mechanistic scheme of things virtually all parapsychological events—telepathy, precognition, psychokinetics—become impossible. The official view on such unaccountable phenomena is what psychologist Alex Comfort, making reference to the great nineteenth century physicist Hermann Helmholtz and thus the whole weight of mechanistic science, calls "the Helmholtzian position": even if it *is* true, I don't believe it. [12] The notion that unlikely coincidences may actually be meaningful becomes absurd. Instead of wondering about their meaning, we simply marvel at the laws of probability, while statisticians rub their hands at the prospect of explaining them away.

But every major mythology, while closing the door on some possibilities, opens others. Mechanistic science is an example of the success of a particular mythic mode of consciousness that has been carried forward with great energy, dedication, and commitment by many people. Examples of other, if radically different, modes that have achieved success in their own spheres of endeavor are shamanism and yoga. Shamans, operating within the frame of their own mythic consciousness, travel to other worlds, act as guides to departing souls at death, and heal the sick with methods that make no sense to physicians trained in American Medical Association–approved medical schools. The yogi, developing a mastery over the inner world comparable to the mastery of modern science over the outer, objective world, may be visited by *siddhis*, or powers, which appear to be miraculous. Neither shaman nor yogi finds reality limited by scientific theories, no matter how those theories might deny their experiences.

The mythos of the mechanistic universe brings our attention like a searchlight onto the stage of sharply defined and causally driven lumps of matter and the forces, fields, and energies that move them. (Physicists call this drama "the particle play.") The new science, however, took up only the notion that the atom is indivisibly small, and applied it only to the external, physical world. The soul was either denied or forgotten. As a consequence, our cultural attention to this play has become narrowed to the point that we have come to overlook the vast penumbra of daily real-world events that surround the stage but are not easily made to participate in the drama. Among these are meaningful coincidences.

The universe depicted by mechanistic science is entirely predictable. Everything is accounted for by laws of causation; there is no "slippage." Newton's equations for the movement of the planets assume a mechanical system of action which, once set into motion, requires nothing further to keep it moving. At first it was assumed that God had started the whole process in the beginning of time, after which it continued on its own like a well-wound clock. Eventually, however, the notion of God seemed unnecessary and was discarded. When Napoleon Bonaparte asked Pierre Laplace why he had not dedicated a mathematical treatise to God, as was customary at the time, Laplace simply replied, "I have no need of that hypothesis."[13]

Ironically, the notions of causation that characterize Newtonian physics grew from much older ideas based essentially in a religious mythos. In the classical Greek and Roman world the Stoic concept of the "iron law of the cosmos" expressed the image of the universe as the actual body of God, in which all parts followed absolute law. Later Christian ideas of divine law owed a large debt to this concept. Such ideas were mainly of religious and political interest during the Middle Ages, but later were applied to the physical world. Descartes, the architect of modern notions of causality, based his thinking expressly on the conviction that God would hold himself absolutely to his own rules.[14] His idea of God's immutable constancy formed the basis of all later notions of causality.

The result was the mechanistic mythos of the Newtonian cosmos. This mythos presents itself in awesome and austere beauty, but at the same time robs us of a sense of wonder about the small events of everyday life. Improbable coincidences are diminished to the trivial. Even major scientific accomplishments are drained of their human, poetic dimensions. Years ago, the novelist Norman Mailer was asked to sit on a television panel during the live broadcast of the first manned landing on the moon. While the rest of the panel talked in a heady fashion of the technological accomplishment represented by the landing, Mailer decried the total lack of poetry in the handling of the affair. An event that from the beginning of time was meant to fill our spirits with wonder and inspiration had been reduced to technological egoism and running descriptions of moon rocks.

Nevertheless, Newtonian physics and the mythos it embodied was spectacularly successful. By the end of the nineteenth century, many physicists had even come to believe that all the basic discoveries had already been made. Some said that the future of physics was simply the adding of more decimal points to existing data. Bright students were sometimes actually discouraged from choosing physics as a career. Lord Kelvin, a prominent physicist of the day, saw only "two small dark clouds" on the horizon. These were minor experimental discrepancies in the mechanical explanations of heat and light.

As the total import of these two dark clouds unfolded in the first decades of this century, they reduced Newtonian physics to a convenient fiction.

THE NEW PHYSICS

During the first decades of this century, the full implications of these experimental discrepancies became apparent in the publication of the general theory of relativity by Albert Einstein and in the publication a few years later of a complete quantum theory by Werner Heisenberg. Taken together, these theories and the experimental work that was to confirm them demonstrated the assumptions of Newtonian physics to be mere approximations of reality. As such they remained useful to engineers, but their mythic power to project maps of reality began to erode. The relativity and quantum theories are now in the process of creating a new mythos, a new topology of reality. This mythos is popularly termed the "new physics," a phrase that refers more to its mythological structure than to any explicit set of suppositions it makes.

According to quantum physicist David Bohm,[15] both relativity and quantum physics share the common perspective of wholeness. Relativity views space not as the void of Democritus—a region of nothingness between solid atoms—but returns us to a vision of the universe as a continuous, unbroken fabric. Atoms are special local characteristics of this fabric. The cosmos is of-a-piece, not empty, but filled with itself, much as a painting is filled with itself. There are foreground and background regions, but the canvas is continuous.

Quantum theory is holistic in quite another sense. It views all action as continuous and unbroken. An experiment involving several atomic particles, for example, is treated as a single whole process. The particles have no individual existence but contribute only to the total event of the experiment. It is the whole situation that the theory deals with, treating the parts as secondary and having no essential substance.

Such notions are more compatible with synchronicity than was the mechanistic model of the cosmos. Synchronicity itself implies wholeness and, therefore, meaningful relationships between causally unconnected events. In quantum theory we recover the view of a world as an unbroken fabric in which seemingly separate events do not occur in isolation but, in fact, form pieces interwoven into a common tapestry. This was Pythagoras' perspective when he looked into the ripple patterns of water to see beyond into the future of human affairs. It is this perspective that was lost in the aftermath of the Newtonian scientific revolution.

The mythic dimensions of relativity and quantum theory, however, are taking surprisingly long to settle in. Discoveries made early in this century are only now beginning to be felt widely. Alex Comfort points out one important reason for this. Unlike Newtonian physics, the new physics gives us little to visualize.[16] We cannot get a clear picture of it in our heads. The cosmos of Newton was easily imagined as a great celestial machine—a cosmic clock—running effortlessly and eternally. The cosmos of relativity is more obscure. It has four dimensions, rather than the three that are obvious to intuition, and time is given an equal status with distance. Only with effort and the aid of projection diagrams can we get much of a picture of this, and then only in an inadequate way. And quantum theory gives us absolutely nothing to hang an optical hat on. Its postulates of probability waves, indeterminacy, and complementarity sound like dialogues from *Through the Looking Glass.*

The pioneers of the old physics believed that they were discovering the divine plan for the cosmos. Their science and their worldview were one and the same. In other words, their science formed a clear mythology. Quantum physics, however, has especially failed to provide such clarity. As Comfort observes, "the revolution involved in quantum physics has had absolutely no impact on the day-to-day worldview even of people who work with it. Unlike the discoverers of the Copernican and Newtonian world these people experience no reordering of consciousness: they say, 'How intensely interesting,' and go home to dinner."[17]

Unlike its classical predecessor, quantum physics presents an open view of the world, one in which the outcomes of events are not entirely predetermined by fixed and inflexible laws. Quantum predictions do not dictate exact experimental results at all, but allow instead for a range of outcomes of differing likelihoods. In this sense quantum theory is *probabilistic*, mapping probabilities rather than specifying events. Some interpreters of quantum physics have bemoaned this uncertainty, feeling that it robs us of the exact knowledge given by Newtonian physics. It rubs the sharp edges off reality, leaving the picture fuzzy. Einstein himself vehemently objected to the probability notion, saying that God does not "play dice" with the universe. (Here Einstein refers quite literally to the idea that natural law is God's adherence to consistent behavior.) Other views, however, are possible. The brilliant systems scientist Erich Jantsch argued that it is precisely this indeterminant character of the quantum world that gives it the openness that was missing in the Newtonian cosmos. The universe at each moment contains the possibility of the unexpected, the new, and even the creative.

CREATIVITY

Nothing is closer to the heart of the experience of synchronicity than the feeling that the world itself expresses creativity in synchronistic coincidences. Such coincidences often have more the feel of poetry than physics. One recalls the beetle arriving at Carl Jung's window during the discussion of the dream beetle. In the case of M. de Fortgibu and the plum pudding, one gets the sense of a clown or trickster standing behind the scenes, the mythic face of a mischievous god, dimly seen, looking from behind the shroud of coincidence. Here we pick up the thread of a major clue to the meaning of synchronicity, the notion that chance may express itself through the mythic theme of a divine Trickster, embodied for example in the Greek god Hermes. We will examine this idea at length in later chapters.

Perhaps the open nature of quantum physics is related to a certain willingness of many quantum physicists to tolerate paradox and ambiguity in their own lives. Niels Bohr proposed the principle of complementarity, by which particles become waves and vice versa, depending upon how they are observed. He carried into his daily life the belief that human situations likewise have opposite and complementary sides. In an interview, he once recalled discovering that one of his children had done something inexcusable. He found himself, however, unable to inflict the appropriate punishment. It was then that he realized that "you cannot know somebody at the same time in the light of love and in the light of justice."[19] Instead, you must choose the context in which you will know another human being: the context of love if you are, for example, a father or mother, or the context of justice if you are, for instance, a judge in a courtroom. You will know one of two different persons depending upon which you choose.

Wolfgang Pauli, like Bohr, was one of the inner circle of scientists who founded quantum physics. He was acutely aware of improbable and creative coincidences in his own life. Arthur Koestler, perhaps the greatest explorer of synchronicity in recent years, gives an example that occurred during a period when Pauli was intensely involved in working on problems related to symmetry in subatomic particles. Pauli visualized much of subatomic activity in terms of mirrors and their reflections, becoming virtually obsessed with mirrors. A friend wrote to him, teasing him about his "mirror complex." But Pauli wrote back, recalling the legend of Perseus and the Medusa. In the legend, Perseus was only able to slay the Medusa (read nuclear physics), whose appearance was so hideous as to turn men to stone, by looking at her reflection in his shield. At about this time Pauli received a paper from a former student, turned biologist, concerning a certain light-sensitive fungus, *mykes* (the Greek word for mushroom). Shortly afterward, Pauli was reading a philosophical essay dedicated to Carl Jung, "about, of all things, the significance of the Perseus legend. It appears that after the Medusa venture, Perseus founded the town of Mykenea, which owes its name to a Greek pun. For on that site Perseus dug up a mushroom; but he had to dig so deep that a brook sprung up from the earth which quenched his thirst. So they called the town Mykene after that mushroom." It was said that Pauli roared with laughter upon reading this. Koestler comments that "this whole tangled web is only a small detail in the cluster of coincidental events which Pauli encountered in critical periods of his life."[20]

Pauli was well known among the physicists of Europe for what was humorously called the "Pauli effect." His presence alone was sufficient to cause complex scientific equipment to misfire. Apparently, theoretical physicists usually cannot handle experimental equipment without breaking it, and Pauli was a very good theoretical physicist! George Gamow, himself a well-known physicist, recounted such an event that did not at first seem connected to Pauli's presence:

> It occurred in Professor J. Franck's laboratory in Göttingen. Early one afternoon, without apparent cause, a complicated apparatus for the study of atomic phenomena collapsed. Franck wrote about this to Pauli at his Zurich address and, after some delay, received an answer in an envelope with a Danish stamp. Pauli wrote that he had gone to visit

Bohr [in Copenhagen] and at the time of the mishap in Franck's laboratory his train was stopped for a few minutes at the Göttingen railroad station.

Gamow comments, "You may believe this anecdote or not, but there are many other observations concerning the reality of the Pauli effect!"[21]

THE SOUL OF CIVILIZATION

Certain prominent themes in the science and culture of our moment in history open the doors of exploration to that which was unthinkable in the Newtonian cosmos. History, says William Irwin Thompson, is the story of the ego of a civilization, while myth is the story of its soul.[22] The ego of today's civilization is still tied to the Newtonian age. The official representatives of technology continue to speak the language of absolute causality. The soul of civilization, however, is changing. The old myths are disintegrating. As to the shape the next mythology will take, the great mythologist Joseph Campbell remarked:

> One cannot predict the next mythology any more than one can predict tonight's dream; for a mythology is not an ideology. It is not something projected from the brain, but something experienced from the heart, from recognitions of identities behind or within the appearances of nature, perceiving with love a "thou" where there would have been otherwise only "it." . . . As stated already centuries ago in the Indian Kena Upanishad: "That which in the lightning flashes forth, makes one blink, and say 'Ah!'—that 'Ah!' refers to divinity."[23]

Campbell speculates, however, that at a minimum, "in the new mythology, which is to be of the whole human race, the old Near Eastern desacralization of nature by way of a doctrine of the Fall will have been rejected." Here, Campbell puts the roots of the notion of a lifeless cosmos in the ancient myth of the fall from Eden, in which only lost Eden itself was sacred territory. In the new mythology, he suggests, all of the cosmos will again be sacred, so that "when lightning flashes, or a setting sun ignites the sky, or a deer is seen standing alerted, the exclamation, 'Ah!' may be uttered as a recognition of divinity."[24]

It may seem from all this that efforts to prefigure the next mythology are bound for failure. Nonetheless, certain broad mythic themes that are important to our topic emerge clearly enough to be outlined. These are discussed briefly in the following pages, though most are not new to us. They include, first, the idea of profound wholeness; second, the notion that at some fundamental level we are all interconnected; third, the concept of the universe as filled with life; and fourth, the idea that creativity is basic to the nature of the cosmos. All are consonant with Campbell's thoughts.

The idea of wholeness, central to modern physics, is emerging as a major mythic theme of our culture. It was first clearly articulated in this century as early as 1926 by the South African statesman and philosopher Jan Smuts in his book *Holism and Evolution*. Smuts viewed the cosmos as formed of wholes, each interlacing with others to form increasingly larger, interconnected tapestries. These structures are not static but evolve toward increasingly inclusive, complex, and even creative forms. This picture of a universe of interwoven and evolving wholes he called "holism," after the Greek *holo*, meaning whole.[25]

One correlate of the theme of holism is the return of the medieval notion that all things are interconnected. This idea finds modern expression in biologist Rupert Sheldrake's theory of morphic fields, networks of resonance that form webs of mutual influence beyond the usual limitations of space and time. We will explore Sheldrake's notions, so reminiscent of the old concept of sympathies, in later chapters. On the level of popular culture, however, the mythic notion of interconnection is seen in the recently fashionable story of the hundredth monkey. This is a loosely reported set of scientific observations which, despite having been widely discredited, has risen to the status of a cultural parable. The story concerns a number of macaques (monkeys) in colonies maintained by the Japanese government. In his book, *Lifetide: The Biology of Consciousness*, biologist

Lyall Watson told of an energetic young female named Imo, who discovered how to clean sand from sweet potatoes by dunking them in a stream. Over a period of time she taught other monkeys this trick, and the skill gradually spread. All at once, however, the rate of spread seemed to undergo a quantum leap and virtually all the monkeys began to do it. Watson wrote:

> Let us say, for argument's sake, that the number [of potato washers] was 99 and that at 11 o'clock on a Tuesday morning, one further convert was added to the fold in the usual way. But the addition of the hundredth monkey apparently carried the number across some critical mass, because by that evening almost everyone in the colony was doing it. Not only that, but the habit seems to have jumped natural barriers and to have appeared spontaneously, in colonies on other islands and on the mainland in a troop in Takasakiyama.[26]

Despite the fact that Watson himself has admitted that this story is not based on sound observation, it has been widely repeated. Its appeal is strong, and it has been told and retold. Its popularity, however, is based not on fact, but on its viability as a parable of wholeness. It tells us again in modern times that we are all connected.

Related to the mythic theme of wholeness is the return of a sense that the universe is filled with life. Smuts's holistic vision of the cosmos is organic, with its emphasis on the evolution of ever-more complex and even creative wholes carrying the sense of a living, developing process. Such a process, of course, is antithetical to the Newtonian vision of a machine that runs relentlessly and unchangingly for all time. Holism is more consonant with the medieval experience of the cosmos, an experience based on the notion that living forces animate all of nature. As late as the early seventeenth century, Johannes Kepler believed that the Earth itself was animated by an inherent spiritual nature. This idea returns to us in our own time as biologist James Lovelock's Gaia hypothesis (after the Greek earth goddess, Gaia) which states that the Earth as a whole exhibits the properties of a living organism.

The return of the sense of life to the cosmos brings with it a return of the possibility of creativity. Smuts's conception of holism included the idea of the new and the unexpected thrown up in the evolution of complex wholes. At the turn of the century, even before Smuts, the French philosopher Henri Bergson had suggested the notion of *emergents* as creative elements that unfold in the process of evolution. As we have seen, a similar idea is embraced by the openness of quantum physics itself, an openness which allows the possibility of the unexpected in each new moment.

David Bohm uses the term *implicate order* to describe a holistic quantum process which underlies ordinary reality as we experience it from day to day. He discussed its implications for wholeness and creativity in an interview:

> The entire notion of wholeness, creative wholeness, is built into the implicate order. . . . It would be similar to the flash of creative insight in our own mental experience. The general tenor of the implicate order implies that what happens in our consciousness and what happens in nature are not fundamentally different in form. Therefore thought and matter have a great similarity of order; we might extend that idea, saying that the creativity and insight that we have may also have its parallel in nature.[27]

In Bohm's connection of mental and physical processes we begin to see the return of a sense of wholeness, life, and creativity to the universe of physics, a sense that brings with it the return of human meaning to the cosmos itself. We begin to recognize ourselves as truly at home in this cosmos. As was true before the rise of mechanistic science, the larger dimensions of the world are again experienced in terms meaningful to the human spirit. The language of this experience is the language of myth, a language that bridges the space between the conscious and the unconscious, a language that speaks of the meaningfulness of events both within the mind and within the greater world in which each of us lives our day-to-day lives. Myth draws us into meaningful relationships with our entire world, an arena that encompasses both our own minds within and the objective events of the world without.

Synchronicity must be understood within this entire context lest it be trivialized into statistical anomalies and stripped of its human meaning. Like myth itself, synchronicity bridges the gaps between the conscious and the unconscious, between the world of mind and the world of objective events. Not surprisingly, synchronicity is therefore ultimately best comprehended in the language of myth.

We will seek our understanding of synchronicity in the pages of science and in the hints of mythic meaning found between the lines on these pages. We will explore various perspectives of contemporary physics, biology, and systems theory with an eye to the literal and an eye to the mythic. We will then look to the wisdom of earlier ages, when myth was the predominant form of human understanding. From this original wellspring we will seek a deeper psychological and spiritual comprehension of synchronicity. We will learn particularly from the Trickster, Hermes in Greek mythology, Mercurius in medieval alchemy, Coyote and others in native American mythology. Of all mythological characters, it is the Trickster who is most associated with chance and synchronicity, who is the bearer of good or ill fortune, who stirs the sands of fate and melds together glad and unhappy chance in patterns guessed only in the gleam of his eye.

part one

SYNCHRONICITY AND SCIENCE

An immense impulse was now given to science and it seemed as if the genius of mankind, long pent up, had at length rushed eagerly upon Nature, and commenced, with one accord, the great work of turning up her hitherto unbroken soil, and exposing her treasures so long concealed. . . . It seemed, too, as if Nature herself seconded the impulse; . . . as if to call attention to her wonders, and signalise the epoch, she displayed the rarest, the most splendid and mysterious, of all astronomical phenomena, the appearance and subsequent total extinction of a new and brilliant fixed star twice within the life time of Galileo himself.

J. F. W. HERSHEL
The Cabinet of Natural Philosophy

· 1 ·

BUS TICKETS:
science discovers synchronicity

---

*We thus arrive at the image of a world-mosaic
or cosmic kaleidoscope, which, in spite of constant shufflings and rearrangements, also takes
care of bringing like and like together.*

PAUL KAMMERER
Das Gesetz der Serie

THE BUS TICKET

The first detailed investigations of
what we today call synchronicity were undertaken early in this century
by the Austrian biologist Paul Kammerer. He approached meaningful
coincidences strictly as objective physical phenomena. He was fascinated by events that repeat themselves in time or space too frequently to be passed off as mere chance. Suppose, for example, that
on a particular day you notice that your bus ticket bears the same
two-digit number as your theater ticket which, in turn, is the same
as the number on the cloakroom ticket given to you later that evening
at a restaurant. Afterward, you go to a party held at a street address
which turns out to be the same number again. This is precisely the
kind of unlikely sequence of events that intrigued Kammerer. In
pursuit of such sequences he spent hours in public places watching
people pass by and noting the incidence of particular hats, articles
of clothing, parcels, and so on. He analyzed the observed sequences

in detail, categorizing them into first, second, third, and higher-order series. Additionally, he developed a complex classification system which emphasized the structural relationships in each coincidence as homologous, analogous, and so on.

Kammerer realized that such coincidences implied an expanded vision of reality. In the fashion of a good experimental biologist in the no-nonsense age of Victorian science, he held the view that such sequences evidenced a previously undiscovered objective principle of nature: a natural physical law which he named the "law of the series."[28] This law expressed a special kind of inertia according to which similar events repeat themselves, spreading like ripples on the surface of water.

Our example beginning with the number on the bus ticket would be considered a series of the fourth "order," because it involved a sequence of four occurrences of the same two-digit number. In addition to sequential recurrences of events Kammerer also compiled series of parallel or concurrent events, those that occur simultaneously or nearly so. In describing the number of such parallel events, he used the term "power."

One of the authors experienced a concurrence of the third power some years ago while he was on the way to pick up a book, *The Psychology of Consciousness*, by Robert Ornstein, previously loaned to a friend. As he started out the door of his office, a colleague engaged him in a conversation about an article in the magazine *Human Nature*, a magazine edited by Robert Ornstein that has since gone out of business. He commented to his colleague that he was, in fact, at that moment on his way to pick up a book written by Ornstein. His attention was then caught by a package that moments before had arrived in the mail. It was from the offices of *Human Nature* magazine. Upon opening it, he found enclosed a copy of *The Psychology of Consciousness*, a gift copy for subscribing to the magazine!

The three events in this third-power set were: first, the fact that he was on his way to pick up Ornstein's book; second, that he was sidetracked into a conversation about the magazine that Ornstein edited; and third, that a copy of Ornstein's book arrived in the mail at that moment. This whole business also enfolded two concurrences of the second power. One was the presence of the book in the conversation with the colleague at the very time of its arrival in the mail. The other was the presence of the magazine in the conversation, corresponding to the arrival of the book from the magazine's offices.

This example gives us a sense of the Gordian knot quality of many such coincidences. Several coincidences seem to be enfolded together so that one does not know just where to start or stop counting them. Kammerer's elaborate sets of categories were intended specifically to handle just this sort of complexity.

The author's experience is similar to many of those reported in Kammerer's book. For the most part they are trivial, though they have a certain quaint appeal that arouses our curiosity. From the age of twenty, Kammerer kept a log of such coincidences which culminated in his book, *Das Gesetz der Serie* (in English, *The Law of the Series*), published in 1919. The book is divided into two major sections. The first is devoted to a meticulous classification of types of coincidences. It includes exactly one hundred cases, beginning with what Koestler calls a "motley collection of instances"[29] under various headings such as names, words, numbers, letters, dreams, and disasters. This is followed by a discussion of the "morphology of series" where Kammerer actually makes the distinction between the order of a series—the number of similar events sequenced in time— and its power—the number that occur more or less at the same time. These are classified according to the number of parameters they share. For instance, the following concurrence, case 10 in his book, involves six parameters. It is a story of two soldiers, both nineteen years of age, both born in Silesia—though they did not know each other— both volunteers in the transport corps, both admitted to the same military hospital in 1915, both victims of pneumonia, and both named Franz Richter.[30]

In the second portion of his book, Kammerer gives a "systematization" of series of coincidences as cyclic, phasic, or alternating; pure or hybrid; homologous or analogous; inverted; and so on. The value of all this fine-grain taxonomy seems questionable. Unlike the biological specimens with which Kammerer was accustomed to working, meaningful coincidences do not seem to lend themselves to definitive categorization. Despite this difficulty, Koestler makes the favorable comment that "however justified skepticism may be, this first attempt at a systematic classification of acausal serial events may perhaps at some future date find unexpected applications."[31]

Kammerer closes the first part of his book by stating, "we have found that the recurrence of identical or similar data in contiguous areas of space or time is a simple empirical fact which has to be accepted and which cannot be explained by coincidence—or rather, which makes coincidence rule to such an extent that the concept of coincidence itself is negated."[32] Here, Kammerer touches on the most essential feature of his study of coincidence: "the recurrence of identical or similar data." In fact, virtually all of Kammerer's cases involve repetitions of the "data." Throughout, he emphasizes the trend for like to attract like. As we have already seen, however, some of the most instructive examples of synchronicity do not display this quality of similarity, but involve rather the coincidence of *meaning*, an idea that we will develop in detail in Part II.

Most of the cases in Kammerer's book are in themselves insignificant, and virtually all fall into the category of curiosities. In spite of this triviality, though, and in spite of his obvious predilection for analysis, Kammerer was able to see through this "motley collection" of synchronistic exotica to a deeper sense of unity, as implied by the quotation at the beginning of this chapter. This must not have been easy, for as Chuang-tzu said many centuries ago, "Tao is obscured when you fix your eye on little segments of existence only."

It is in the second major part of "the law of the series" that Kammerer gets down to theoretical business. Here he develops his ideas regarding the nature of coincidence. Like Carl Jung and Wolfgang Pauli, who were to follow, Kammerer postulated an acausal organizing principle which he placed on a par with causality itself. This acausal factor draws like and like together. It is a special kind of gravity that operates in terms of form rather than in terms of traditional physical variables such as mass and energy. Events similar in form mutually attract each other to yield recognizable series of events.

Kammerer believed that clusters of recurrent events propagate in cyclic fashion, like waves of water. Only the peaks of the waves are visible; the troughs are hidden from view. Thus, for example, gamblers experience a run of good luck at each peak. At other times they do not do so well. In support of this view, Kammerer reviewed previous theories of periodicity. These included the Pythagorean symbolic use of the number 7 (for example, there are seven musical notes in an octave and then they repeat); Goethe's "revolving good and bad days"; and Freud's belief in twenty-three- and twenty-seven-day cycles that combine to affect behavior. As a biologist, Kammerer felt that cyclic behavior was characteristic of natural processes in general.

A major sector of modern systems theory known as chaos theory or chaos dynamics is devoted in part to just those types of periodic processes that so fascinated Kammerer. *Turbulent Mirror* by John Briggs and David Peat is a highly readable introduction to this mathematical approach to processes that appear random on the surface but on careful scrutiny are found to follow certain general patterns specified by mathematically defined *attractors*.[33] Some of these, known as "chaotic" or "strange attractors," depict processes that are entirely unpredictable in the short run but in the long run show periodic fluctuations. An example is the weather. Short-range predictions of the weather are, as we all know, notoriously unreliable. This is because of the large number of factors that interact to affect it and also because even the smallest of factors may exercise inordinately large influences on the final result. One day it rains; the next day is dry. On the other hand, long periods of wet weather that last weeks, months, or even years are followed by similarly long periods of dry weather. Luck in gambling seems to follow a similar course. Periods of good luck are followed by periods of bad luck, though gamblers do not always win during streaks of good luck, nor do they always lose during streaks of bad luck. It is worth noting that the founders of chaos theory were more than casually intrigued by the behavior of the roulette wheel. Sessions with the computer were sometimes followed by trips to casinos for field tests!

Chaos theory does not *explain* processes such as the weather or the behavior of the roulette wheel, but it does demonstrate that mathematical analysis is capable of describing seemingly random processes that look very much like those which interested Paul Kammerer. Many of these, like the weather, are driven by natural forces that have little to do with the unique quality of personal significance so characteristic of the instances of synchronicity that would later engage and hold the attention of Carl Jung.

Even without the benefit of modern chaos theory, however, Kammerer's cases, virtually all of which involved repetitions of one sort or another, were more than a little vulnerable to explanation as chance alone—or perhaps we should say to the complex interaction of unknown factors that have little or no significance for the individual. Although, as mathematician Michael Shallis clearly points out, probability and statistics have at best a questionable application in such matters, each of us is nonetheless bound to experience a certain number of improbable combinations of events.[34] To make too much of these might be like watching clouds and making a fanfare over those few that have remarkable shapes. Both Shallis and Jung criticize Kammerer on such grounds, though Einstein commented favorably on his book, saying that it was "original and by no means absurd." (See Appendix II for an in-depth comparison of probability and synchronicity.)

It was Paul Kammerer's fate to be criticized. He was one of the last Lamarckian evolutionists. These biologists believed that the individual experience of an organism could influence its offspring. A standard example is that a lion that develops great running speed by virtue of its own unique experience might pass some of this acquired skill along to its cubs. Such notions were in stark conflict with the dominant mechanistic biology of the early twentieth century and put Kammerer in direct conflict with the leading biologists of his day. One of his chief opponents in battles of theory was the great Darwinian William Bateson, father of Gregory Bateson the well-known biologist, philosopher, and founder of the double-bind theory of schizophrenia.

In seeking support for his Lamarckian ideas, Kammerer had acquired a remarkable knack for experimentation with reptiles and amphibians. His specimens, however, were lost during World War I, and the last one preserved, a "midwife toad" (*Alytes obstetricous*) was found to have been tampered with. In September 1926, personally humiliated and his reputation in shambles, Kammerer took a stroll up an Austrian mountain path and shot himself in the head. His life's story is told in a fascinating biography written by Arthur Koestler, *The Case of the Midwife Toad*.[35] This book contains an appendix, "The Law of Seriality," which is the best introduction to Kammerer's ideas on synchronicity available in the English language (*Das Gesetz der Serie* has not been translated).

2

SYNCHRONICITY IN THE HOUSE OF PHYSICS

First he conceived from the depth of his being a something, neither mind nor matter, but rich in potentiality. . . . It was a medium in which the one and the many demanded to be most subtly dependent upon one another; in which all parts and all characters must pervade and be pervaded by all other parts and all other characters; in which each thing must seemingly be but an influence in all other things; and yet the whole must be no other than the sum of all of its parts, and each part an all-pervading determination of the whole. It was a cosmical substance in which any individual spirit must be, mysteriously, at once an absolute self and a mere figment of the whole.

OLAF STAPLEDON
Star Maker

As we saw in Chapter 1, Paul Kammerer tried in his own brilliant and unorthodox way to put coincidence under the lens of objective science. The physics of his day, however, and the mythos that it represented, gave little support for any effort to go beyond its own mechanistic notions of causality. It had, to use Alex Comfort's term, little *empathy*[36] with synchronicity. Steeped in this mythos, Kammerer's final formulation of the law of the series amounted to little more than a vague notion about the replication of similar events. It is our intention in this chapter to move beyond Kammerer and to reconsider synchronicity in the light of a contemporary physics, which breaks the mold of mechanistic explanation and is therefore more empathetic with the unexpected and the creative.

THE UNBROKEN PROCESS

Synchronicity implies a cosmos in which seemingly unrelated events are woven together to form a continuous world fabric. Such a cosmos does not square with the classical, mechanistic physics that views the universe as a loose assemblage of objects, forces, and energy.

At the beginning of the twentieth century the house of classical physics was built of the strongest stuff imaginable—the impenetrable atoms which were the basis of the physical world. The location of each of these atoms was precisely fixed in the three-dimensional space of Descartes' geometry. In a very few years, however, the house of physics was shaken to its very foundations. The atoms of its walls dissolved into abstractions, and these in turn became probabilities written on the blackboards of mathematicians. Its location became indeterminant. The very space it occupied became curved, warped, and even filled with "worm holes"! It was assaulted by Einstein's general theory of relativity on the one hand and quantum theory on the other. Both new theories view the cosmos as an undivided whole, though they do so in very different ways.

Beginning with Einstein the entire cosmos becomes an undivided field. Objects such as atoms and stars are viewed as properties of this field. They are seen as local concentrations of it. Think of vortices on the surface of water. Each consists of a stable, whirling movement which gives a unique form to that particular part of the surface. While such vortices can interact with each other, combining to form new and larger vortices or canceling each other out, they have no separate existence in and of themselves but are simply local characteristics of the water.

Unlike the general theory of relativity, quantum theory does not deal with the existence of objects but with actions or events. It is an elaborate mathematical structure within which separate objects have, in fact, no representation. Only events have reality for the theory, and these are all intimately interconnected. For example, if several subatomic particles are involved in an experiment, they cannot be treated as separate realities. The outcome of the experiment is determined by the total quantum state of the system of particles, which cannot be viewed in any real sense as comprised of separate and independent objects. Instead, the mathematics of quantum physics treats any system of particles as part of a larger unity.

Wholeness in quantum physics even includes the experimental apparatus by which measurements are made. Beyond this, it includes the entire context of the experiment: the laboratory, the experimenters themselves, and so on. Quantum physicist David Bohm has observed, "Ultimately, the entire universe (with all its *particles*, including those constituting human beings, their laboratories, observing instruments, etc.) has to be understood as a single undivided whole, in which analysis into separately and independently existent parts has no fundamental status."[37]

MICROWORLD SYNCHRONICITY

The wholeness suggested by synchronicity lies in meaningful connections of events isolated in time and space; in spite of their separation, they seem to be linked together. Something like this is known in quantum physics, and perhaps not surprisingly it is referred to there as synchronicity.

The story of quantum synchronicity, like many stories in modern physics, begins with Einstein. Though he was one of the pioneers of quantum physics, he was never a friend of it, especially as it was interpreted by two of its founders, Niels Bohr and Werner Heisenberg. One outcome of Heisenberg's equations was the realization that the amount of information we can obtain about a particular atomic particle is limited. Put differently, we can reach only a limited degree of precision in describing such a particle. For instance, we can measure the position of a particle at any instant in time with as much precision as we like, but the result will be that the velocity (technically the momentum) of the particle will become unknown. In fact, we can measure the position *or* the velocity with as much accuracy as we like, but in obtaining the one we lose the other. Bohr interpreted this "indeterminacy" as meaning that the measurement of, say, velocity leaves position indeterminant in a fundamental way: Position no longer has any meaning whatsoever. We can put position in the same class as the sound of one hand clapping: It is fascinating to think about but has no object of reference in the world of reality.

Einstein was unimpressed with all of this, and over the years he proposed a number of "thought experiments" designed to demonstrate that Bohr's thinking was incorrect. A thought experiment is an imaginary experiment designed to demonstrate that some aspect of a theory succeeds or fails on logical grounds alone. Einstein published the most famous of these in 1935 with Boris Podolsky and John Rosen; it has since become known as the EPR paradox.[38] Einstein, Podolsky, and Rosen built their experiment upon certain subtle facts that follow from the mathematics of quantum theory. In particular, two particles that share a condition known as a *singlet state* (for example, they may have been two halves of some larger particle), retain a special relationship with each other even after they separate and go their separate ways.

If, for example, the original larger particle splits, ejecting the two component particles in opposite directions, the positions and velocities of the two remain correlated, though one travels, say, to New York and the other to Paris. Knowing the position of one allows us to specify the position of the other. The same is true of velocity. An example often used is the interaction of two billiard balls. Suppose that two such balls interact on the pool table: one strikes the other. At the instant of impact they exist as a single "particle" which immediately disintegrates, sending its component parts in opposite directions. If we measure the position of one we can infer the position of the other. Similarly, if we measure the velocity of one we can infer the velocity of the other. The positions of the two balls are precisely correlated, as are the velocities.

If these were quantum billiard balls, however, they would have a very unusual property: measuring the position of one of them would make it impossible to measure its velocity also. This is Heisenberg's limitation. Einstein, Podolsky, and Rosen simply said, why not measure the position of one and the velocity of the other? From these two measurements and the correlation of the balls' positions and velocities, we can know both measurements for each ball: we can have complete knowledge of them.

Einstein wanted to show first of all that Heisenberg's constraint on complete knowledge was artificial. But more importantly, he wanted to demonstrate that quantum theory is an incomplete description of reality. But consider Bohr's reply. In essence he stated that the EPR paradox artificially separated the measurement of the two particles. Making a measurement on one particle, he said, would effectively "blur the frame of reference" of the other. Bohr's reply implied that the particles were, in fact, no more independent after leaving the singlet state than they were before. This idea is holistic in that it emphasizes the notion that the two particles cannot be treated as separate objects even though they occupy different locations in space and may, in fact, be quite far from each other.

This remarkably holistic feature of Bohr's reply to the EPR paradox was highlighted in 1965 by the British physicist John Bell, who restated Bohr's ideas in a form testable in the laboratory.[39] He proposed a set of mathematical proofs, now famous as Bell's inequalities, which demonstrate that a pair of particles, once in the singlet state, retain an interconnection despite their different future locations in space. Bell's demonstration of interconnection at a distance shifted the emphasis of Bohr's original argument toward the holistic implication of *nonlocality*—that particles need not share the same local region of space to be intimately interconnected.

Bell's inequalities deal with the "spin" of atomic particles, rather than their positions and velocities. (Spin is a formal quantum characteristic which corresponds roughly to the rotation of a billiard ball.) The inequalities predict that the laboratory measurement of the spins of separate particles which were once united will be correlated to a significantly higher degree than would be expected on the basis of classical physics. To date several experiments have been conducted, and there seems to be little doubt that the inequalities are correct. The spins of such particles are indeed correlated at a distance, apparently through no causal mechanism allowed by classical, mechanistic physics. In quantum physics, this correlation is termed "synchronicity." The full implications are succinctly stated by physicist Nick Herbert, who observed, "A universe that displays *local phenomena* built on a *nonlocal reality* is the only sort of world consistent with known facts and Bell's proof."[40]

It has been said that if you are not shocked by quantum physics, you simply do not understand it. Certainly Bell's inequalities sent a shock through the house of physics. One way that physicists deal with the shock is to treat the inequalities simply as a set of mathematical predictions and nothing more. This, in fact, has been a common attitude, but it has not proved very satisfying. An alternate view, developed by David Bohm, proposes that the two particles are not separate at all but, instead, constitute two different views of one single particle.[41]

Suppose, says Bohm, that this single particle were a fish in a tank, with two video cameras set up to view it through two separate monitor screens. Put the cameras at right angles to each other, one in front of the tank and the other beside it. What we see on the monitors appears to be two separate but very remarkable fish, because for every motion that one makes, a correlated motion will be observed from the other. Not realizing that the images on the monitors are the same fish, we might conclude that they are separate, connected to one another by some unknown causal chain. Bohm says that we see each of the fish (read particles) separately in three-dimensional space, while in reality there is only one fish (particle) which exists in six-dimensional space. The two are the same when seen from this higher dimensional perspective.

Now, the above observations deal with the "microworld" of ultimate particles. It seems to us, however, that we are dealing with a similar phenomenon in the synchronicity experienced in the "middle world" of ordinary day-to-day living. In both instances we are faced with highly correlated sets of events for which there is no causal explanation. In both instances what we actually see appears to be no more than the surface display of a deeper, unseen event connecting the separate parts.

Middle-world synchronicity, however, seems more complex than its microworld counterpart. In the latter, the two separately appearing particles are no more than mirror images of each other, whereas in middle-world synchronicity we seem to see two distinctly different facets of a larger, hidden pattern. Here, the surface display that we actually see might better be likened to something seen on a video-game screen. The actions of the various "objects" on the screen are highly correlated with each other but do not represent manifold reflections of the same image. Rather, they are all generated from a larger, implicit pattern unfolding in the circuitry of the video-game's computer. It is this hidden pattern that produces the seen events that fit together so smoothly. As we will soon see, this metaphor, originally suggested by Alex Comfort,[42] is an apt description of an all-encompassing cosmological model that David Bohm has recently been developing.[43]

THE HOLOGRAPHIC ORDER

The modern physics worldview that is most empathetic with synchronicity is the holographic order now being developed by Bohm. Unlike the cosmos inherited from Descartes, which can be imagined as a topological map of objects and events occupying separate locations in a three-dimensional coordinate system, Bohm envisions the cosmos as a hologram. His vision is radically holistic, allowing for the creation of separate but correlated events beyond the bounds of causality.

A hologram is made from a holographic plate. When viewed under ordinary light, such a plate looks much like an underexposed photographic negative. When viewed under special lighting conditions, such as laser light, it takes on the appearance of an open window. Just as through a real window we can see the entire scene beyond by looking through any part of it, in a hologram the whole landscape is visually contained, or, to use Bohm's term, *enfolded* into each part. The essential wholeness of the hologram is that each part contains, or enfolds, the whole. On the cosmic scale of Bohm's theory, this means that each part of the world contains the whole of the universe hidden (enfolded) within it. This stunning notion, though new to modern science, is well known in the world's mystical poetry. In *The Garden of Mystery*, the Sufi mystic Mahmud Shabistari states:

> Know that the world is a mirror from head to foot,
> In every atom are a hundred blazing suns.
> If you cleave the heart of one drop of water,
> A hundred pure oceans emerge from it. . . .
> In the pupil of the eye is a heaven.
> What though the corn grain of the heart be small
> It is a station of the Lord of both worlds to dwell therein.[44]

A hologram is created when a pattern of light-wave interactions is captured on the holographic plate. The capacity to enfold large, whole images within small parts seems to be characteristic of such patterns of interactions. An example is the pattern of ripples seen on the surface of a pond a few seconds after you toss in some pebbles. These ripples create complex figures as they expand, crisscrossing over the surface of the water, each spreading from its own source where a pebble fell. If we could freeze the pond instantaneously, these ripple patterns would contain the information necessary to reverse the process, recreating the original configuration of pebbles as they struck the surface. The configuration of falling pebbles, we might say, is enfolded by the patterns of the ripples. This is exactly what a hologram does, using light waves instead of water ripples.

On a grand scale, we might envision the entire cosmos as a vast light-pond with ripples spreading, interpenetrating, and creating complex patterns of interactions throughout. Certain of these patterns may seem relatively stable, others may not, or they may give the appearance of stable configurations in motion. This is the vision of the *holomovement* at the heart of Bohm's concept of a holographic universe.

Normally, says Bohm, we think of the universe as comprised of more-or-less solid objects such as atoms or stars, many of which emit light and other forms of electromagnetic radiation (radio waves, gamma waves, and so on). The space between objects is filled with a constant flow of radiation. Ordinarily, we do not take much account of this ocean of electromagnetic waves. Instead, we consider solid objects to be real and the flux of energy to be of secondary significance. But Bohm flips this picture upside down, making flux the primary reality of the cosmos. He views solid objects as stationary patterns of wave interactions that emerge out of the holomovement rather than points of primary reality. Most of us have seen analogous examples of wave interactions in the tiny stationary patterns of ripples on the surface of a cup of coffee or tea when the table is set into vibration, perhaps by a large motor running nearby. These small "standing wave" formations are the product of a less apparent but more basic action of ripples flowing swiftly about the surface of the cup, reflecting off its walls, and interacting with each other. This is analogous to Bohm's notion of how solid matter is generated out of the flow of electromagnetic waves. As Einstein in the general theory of relativity had viewed objects to be stable configurations of the space-time continuum, so Bohm views objects as stable patterns of motion. With this vision of reality, Bohm hopes to unite relativity with the other great cornerstone of modern physics, quantum theory.

The universe according to Bohm actually has two faces, or more precisely, two orders. One is the *explicate order*, corresponding to the physical world as we know it in day-to-day reality, the other a deeper, more fundamental order which Bohm calls the *implicate order*. The implicate order is the vast holomovement. We see only the surface of this movement as it presents or "explicates" itself from moment to moment in time and space. What we see in the world— the explicate order—is no more than the surface of the implicate order as it unfolds. Time and space are themselves the modes or forms of the unfolding process. They are like the screen on the video game. The displays on the screen may seem to interact directly with each other but, in fact, their interaction merely reflects what the game computer is doing. The rules which govern the operation of the computer are, of course, different from those that govern the behavior of the figures displayed on the screen. Moreover, like the implicate order of Bohm's model, the computer might be capable of many operations that are in no way apparent upon examination of the game itself as it progresses on the screen.

The notion of an implicate order is important to our understanding of synchronicity because it shows us that the cosmos contains possibilities different from and greater than we had suspected. Descartes' universe, for instance, is limited by the idea of *locality*: events must be near one another to be related. But Bohm's picture of the implicate order is that of a vast, scintillating hologram. As such, it shares an important feature with all holograms, one with which we are already familiar: Each part contains or enfolds the whole. Locality does not have primary significance here, as the whole is entirely enfolded in each part. Locality is a property of the explicate order and not of the deeper implicate order from which the explicate order unfolds.

In our ordinary ways of seeing, we have perhaps overrated the importance of locality as something of primary significance in and of itself. Bell's inequalities would suggest this. Like many other regularities observed in the natural world, locality may be important under certain fixed conditions but not under others. For example, the appearance of meaningful patterns of events such as those seen in synchronicity—patterns which are not connected by local chains of causality—become possible in the holographic worldview.

According to Bohm, what is more important than locality is the *degree of enfoldment* shared by apparently separate events. In this way events separated in space or time can share a more essential relationship than other events which may be physically near each other. Two persons who even at a distance share a common thought or feeling have more in common in this regard than do two who are near each other, who are even perhaps engaged in conversation with each other but do not share a common state of mind. Notions such as these, which run directly against the grain of mechanistic science, are strongly empathetic with synchronicity. Indeed, as we know, synchronicity often involves a coincidence of separate events that seem connected in a way that makes sense only outside the usual limited ideas of causality.

THE PATTERN THAT REPEATS

Before we return to Bohm's ideas regarding the holographic universe, let us take a side excursion to examine some coincidences that involve repetition. We will meet Rupert Sheldrake, a British biochemist whose ideas mesh with synchronicity and the holographic perspective, and Ervin Laszlo, a Hungarian systems theorist who has developed similar ideas in terms of quantum physics.

Perhaps the most common type of synchronistic experience, and the type Kammerer focused on almost exclusively, involves an idea, theme, or pattern of events that repeats itself. It can be a name, a number, a topic of conversation, an item of clothing, or just about anything. Recently, for example, one of the authors was talking with a friend about Sheldrake's theories, discussing possible relationships to Bohm's notion of the implicate order. He happened to have a written transcript of a conversation between Sheldrake and David Bohm, which he loaned to this friend. The latter read it while listening to the local public radio station. The station happened to be airing one of a series of interviews titled *Physics and Beyond*, and he found himself listening to an interview with David Bohm in which he was talking about Sheldrake's theory!

A variation on this theme involves the actual appearance of something that previously existed only as a thought or idea. The arrival of the scarab beetle at the time Jung's patient was relating her dream of this insect is a case in point. So are many instances of Arthur Koestler's "library angel," his tongue-in-cheek spirit associated with lucky coincidences involving libraries, quotations, references, and the like.[45] An excellent example of the angel at work was reported to Koestler in 1972 by Dame Rebecca West, who was researching a specific episode that took place during the Nuremberg war crimes trials:

> I looked up the trials in the library and was horrified to find they are published in a form almost useless to the researcher. They are abstracts, and are catalogued under arbitrary headings. After hours of search I went along the line of shelves to an assistant librarian and said: "I can't find it, there's no clue, it may be in any of these volumes." I put my hand on one volume and took it out and carelessly looked at it, and it was not only the right volume, but I had opened it at the right page.[46]

Koestler comments that coincidences of the library angel-type are "so frequent that one almost regards them as one's due."

An instance of repetition synchronicity which involved three elements (a third-order coincidence, according to Kammerer's classification) occurred to one of the authors while he was driving across town. He noticed that the car radio was playing an old popular song about "bad, bad, Leroy Brown," who among other things is said to be "meaner than a junkyard dog." The phrase stuck in his mind. He imagined that there must actually be such dogs living out their lives in junkyards, growing older and meaner by the day. These reflections were cut off abruptly as he switched stations. Instantly, he heard an advertisement for a local junkyard billing itself as the "home of the junkyard dog." He happened to look up and notice that he was passing a large junkyard. The sign read, "Home of the Junkyard Dog"!

These cases involve common themes or ideas that kept cropping up, either in written form, in thought, or in concrete reality. Quite simply, there is in each instance a pattern that repeats.

Formative Causation

Rupert Sheldrake is a biochemist with a special interest in how different species of organisms develop, each into its own unique and characteristic shape or form. This is the study of *morphogenesis* or "the coming-into-being of characteristic and specific form in organisms." In recent years morphogenesis, like much of biology, has been dominated by an analytic and reductive approach to living systems, viewing them fundamentally as biochemical systems to be studied on the molecular level. This approach has yielded some spectacular advances, such as the formulation of the structure of the genetic code in the DNA molecule. But Sheldrake points out in his book, *A New Science of Life*, that the purely biochemical approach does not now, nor is it likely to in the future, provide a complete explanation of morphogenesis. He supplements the molecular theories with a holistic proposal, his hypothesis of formative causation. [47]

Sheldrake's central idea is that the development of a living organism is controlled by a kind of holistic field or force. Such a notion is not new. The idea of an overarching principle of formation can be traced back to Plato's "ideal forms," which existed in a higher reality of their own, serving as models for the less-than-perfect forms of this world. The early twentieth-century vitalists, especially Hans Driesch and Henri Bergson, also argued that a living organism is more than a physical assemblage of molecules. There is an overarching holistic principle (for instance, the famous *élan vital* of Bergson) which gives its development an overall direction and integration. This principle fell into disrepute in the mid-twentieth century, both because of the prevalent mechanistic philosophies of the time and because of the successes of the molecular approach.

A difficulty with the vitalist's ideas, as well as with Plato's ideal forms, is that these high-order formative, principles have a rigid or static quality which is inconsistent with the evolutionary change that is so much a part of nature. Sheldrake's proposal of a formative field, the *morphic field*, however, provides an overall patterning of form which is at the same time subject to change.

A morphic field is a kind of habit of nature. Each time a particular form occurs, it is more likely to occur again, whether nonorganic, such as an atom, a molecule, or a snowflake, or living, such as a flower, a bird, or a human. Sheldrake also believes that morphic fields influence patterns of brain activity associated with thought or behavior.

In embryological development, the morphic field acts on the DNA molecule much like a radio wave acts on a radio, giving its output a specific form without actually altering its hardware. An important aspect of the radio wave is that it supplies very little of the actual energy needed to produce sound from the radio. Rather, the wave supplies a minute amount of energy patterned in such a way as to guide and structure a final output that itself may involve considerable amounts of energy. Likewise, morphic fields require minute energy to exert dramatic influences upon nature. This seems at first a strange idea, but many processes in nature start on a micro scale that readily may be influenced by the minutest quantities of energy. Consider the difference between growing roses and lilies. During the early embryonic molecular events the smallest possible forces could push the ensuing development toward one or the other. At this stage it is not a matter of energy so much as of information. The genetic code for the rose contains different information than that of the lily and represents, Sheldrake would say, a different morphic field. A similar case can be made for electrical activity in the brain, which also starts at the minutest levels of energy and develops into processes that involve large areas of the nervous system. Indeed, the nervous system is a natural place to look for the subtly whispered influences of morphic fields.

One morphic field that exerts its influence on the nervous system is called a "motor field." Motor fields may be important in producing genetically programmed behaviors, such as the tendency of small animals to run for cover when they see the shadow of a hawk. Motor fields may also provide a new model for explaining learning and memory; that is, memories are equivalent to motor fields built up from past experience. Morphic fields have a quality we have already seen in quantum physics: they are not limited by location. This means that, though a person's individual memories must somehow be matched to the pattern of his or her own unique nervous system, one person's experience can influence others. In effect, when something is learned once, it is more easily learned again later by someone else. A pattern of thought or behavior is more easily produced once it has been produced before. This theory, interestingly, gives the first scientifically reasonable account of Jung's notion of psychological *archetypes*, universal images or themes shared by all humanity. Jung himself believed that archetypes are built up across aeons of historical time, an idea that could not agree more with the way morphic fields are said to be formed. Sheldrake recently observed, "Morphic resonance theory would lead to a radical reaffirmation of Jung's concept of the collective unconscious," that is, of archetypes.[48] We will examine archetypes in detail in Chapter 4.

We can see a linkage between morphic fields and synchronicity in the often-celebrated fact that two or more scientists or mathematicians may make very similar and independent discoveries at almost the same time. An excellent example is the calculus, developed in England by Sir Isaac Newton and almost simultaneously in Germany by the philosopher, scientist, and mathematician G. W. Leibnitz. Newton knew nothing of Leibnitz's work and, in fact, made do with a considerably more awkward mathematical format. (It is Leibnitz's form that is used today, though Newton is usually given credit for the method.) Such coincidences are often attributed to cultural conditions: everything was just right for the discovery. In many instances this is without doubt true, but in some cases it seems a less probable explanation, as we shall see further on in the chapter as we discuss the independent publication in the same year of three theories, including Sheldrake's own, each of which postulates a tendency for patterns, once created, to reproduce themselves.

Other examples of meaningful coincidences which we may explain by the existence of morphic fields are more common but less dramatic. They include relatively frequent situations in which two or more persons are thinking or doing similar things at the same time but with no knowledge of each other. For instance, you receive a call from a friend just when you are thinking of calling her; both of you were contemplating the conversation prior to the call. Or you find yourself thinking of something just when a person nearby starts to talk about it, as if to spare you the trouble! In instances such as these, synchronicity may overlap with what we normally think of as telepathy. We will comment briefly on this connection further on. For the moment, though, let us return to our discussion of morphic fields and memory.

Scientific evidence for the presence of morphic fields in memory is sparse but engaging. In the first quarter of this century, one of the founders of American psychology, William McDougall at Harvard University, discovered by chance that untrained rats were exceptionally quick to learn a task (escaping from a water maze) previously mastered by many earlier generations of rats of the same strain. His findings were strikingly confirmed some years later by researchers in Scotland and Australia, where previously untrained rats picked up the task almost immediately.[49] At about the same time the great Russian physiologist Ivan Pavlov—famous for his studies of conditioned reflexes in dogs—made similar observations on several generations of mice trained to run to a feeding place at the sound of a bell. The first generation required an average of three hundred trials to learn the task. The second generation required only about one hundred trials, while the third and forth generations learned in thirty and ten trials respectively. Pavlov found it difficult to replicate these findings. This is not surprising, given the hypothesis of morphic fields, because mice in later studies would benefit by the previous learning of those in the original study, as indeed was found by the Scottish and Australian researchers who tried to replicate William McDougall's findings.[50]

Frequently cited in support of Sheldrake's theory is the story of the hundredth monkey described in the Introduction. Unfortunately, we have at best only the most informal reports of the phenomenon and these do not support it as factual. But the story carries a powerful mythic message of connection. A reliable instance of a similar nature is chronicled in Sheldrake's recent book, *The Presence of the Past*,[51] and concerns the well-documented spread of a simple learned behavior in a small bird, the British blue tit. A small number of these birds evidently learned to open bottles of milk delivered to people's homes by pecking a small hole in the cap and tearing the foil back to drink the milk. They might drink as much as two inches of milk, and occasionally one was found drowned in the bottle! There have been reports, as well, of blue tits following delivery trucks and breaking into bottles while the driver made deliveries.

This activity was first reported in Southampton, England, in 1921, and its spread was recorded at regular intervals through 1947 when it could be seen in many locations in England, Scotland, and Ireland, as well as in Holland, Denmark, and Sweden. While a conventional explanation is possible along the lines of simple imitation, certain facts argue in favor of the active role of morphic fields in the spread of this behavior. First, blue tits are birds that do not usually travel far from their breeding place, while the habit of opening milk bottles appeared at a number of locations many miles from previous citings, including the spread to the Continent. Sheldrake estimates that the habit must have been rediscovered independently at least eighty-nine times in the British Isles alone. Moreover, as the habit was practiced by increasing numbers of birds, it spread with increasing speed, suggesting the buildup of a powerful motor field for this behavior. A particularly instructive instance of its spread was seen in Denmark, where milk bottles almost disappeared during World War II but reappeared in 1947 and 1948. Few if any blue tits would have lived long enough to carry the habit forward from the prewar years, yet it reappeared rapidly when milk bottles again became available.

In his book, Sheldrake reviews a considerable mass of evidence in support of morphic fields, including a number of fascinating anec-dotal cases. Among these are several instances of learned behaviors that animals apparently passed on to their offspring. One such case, originally recorded by Charles Darwin, involved a mastiff that, due to mistreatment at the hands of a butcher, developed a strong dislike for butchers and butcher shops that carried through at least to the second generation of its offspring.

In a recent experimental test of the hypothesis of formative caus-ation, Sheldrake arranged for a picture containing the hidden face of a Cossack to be presented on British television. While viewers watched, the facial features, complete with handlebar mustache, gradually emerged from a puzzle background. Groups in Europe, Africa, and America were then shown the picture. Their ability to recognize the hidden figure increased dramatically after it had been viewed by the British audience. Sheldrake suggests that these later viewers were tapping into a morphic field that had been created by the initial audience.

In a related study, Yale University psychologist Gary Schwartz[53] presented students with a large number of Hebrew words from the Old Testament. He presented some of the words as they are normally printed and randomly scrambled the letters of others. Students who did not know Hebrew guessed at the meanings of the words, indicating their confidence in each guess. Schwartz found, as Sheldrake's theory would predict, that the students rated the real words with considerably greater confidence than the ones that had been scrambled (though they did not accurately guess their meaning). Moreover, he found that confidence ratings were about twice as high for the words that occur frequently in the Old Testament compared with those that occur only rarely. The idea here is that the real words had, in fact, been learned by countless persons throughout history, forming strong morphic fields; the most frequently occurring words, of course, had been seen and read the greatest number of times. The possibility that the real words were simply more easily grasped was eliminated by the ratings of linguistic psychologists, who found the scrambled words to be as structurally sound as the real ones. Similar experiments have been carried out using Persian words and even Morse code.

There is much in Sheldrake's theory that makes it compatible with Bohm's notion of the implicate order. Both Bohm and Sheldrake are holistic in their approach, and both theories assume nonlocality. Is it reasonable to think of Sheldrake's morphic fields as a feature of Bohm's implicate order? One of the authors of this book met Sheldrake in Bombay, in the winter of 1982. He had not yet met Bohm and was less than eager to see his theory framed in the context of Bohm's holographic model. Sheldrake had not explored the model in depth and was concerned that it represented another form of the mechanistic approach. Since then he and Bohm have met and have discovered that they have much in common.

Talking with Sheldrake, Bohm observed that the concept of morphic fields has many of the properties that Bohm has proposed in his concept of quantum potential.[54] This idea has its roots in an earlier notion suggested by the French quantum physics pioneer Louis de Broglie, who in 1927 proposed that individual particles such as electrons are guided or directed by "guide waves." The proposal was not well accepted at that time. In the 1950s, however, Bohm conceived a similar idea in the form of the quantum potential and worked with de Broglie to develop it. More recently, Bohm has again become interested in this concept. He points out that the quantum potential has many of the properties of morphic fields. Its effect is nonlocal. It "guides" the particle in a fashion analogous to the way a radio signal guides an airplane or a ship, by supplying information rather than energy. Furthermore, it is holistic in the sense that it is the product of the entire situation in which it occurs. Of course, the morphic field must direct a great deal more than the flight of a single particle. It must direct the vastly more complex development of an entire organic structure, pattern of behavior, or memory. But the idea is the same.

This problem of spanning the considerable gap between the microworld of the quantum potential and the macroworld of morphic fields and synchronicity may be solved within the brilliant theoretical work of systems theorist Ervin Laszlo.[55] His psi-field hypothesis postulates that mathematical wave functions equivalent to Bohm's quantum potential are built up into increasingly higher-order or "nested" structures that have direct influence upon complex real-world events. These structures or patterns are retained in nonlocal psi-fields analogous to Sheldrake's morphic fields. The two theories are not the same, because the psi-field hypothesis deals explicitly at quantum-level reality. Moreover, its main purpose is to explain a broad range of phenomena, especially certain aspects of organic evolution, which do not interest us here. However, Laszlo seems to have made deep inroads into the problem of translating microworld events into the macroworld events of everyday life. In doing so, he has postulated that patterns of events once created are not lost and are more likely to occur again. His ideas are exactly what is needed for a perspective of the physical world that is empathetic with the observed facts of synchronicity.

We are still confronted with the observation that synchronistic coincidences often are a great deal more than simple repetitions. Bohm states that as the implicate order unfolds to manifest the moment-to-moment reality of the explicate order, new "creative" patterns tend to emerge.[56] These patterns are expressions of a creative urge in the deepest implicate aspect of the cosmos. Such patterns have direct expression in the middle world of everyday life and are not limited by the constraints of locality. Examples include the creation of both biological life and of consciousness. In Bohm's view, these do not occur simply as a result of unique happenstance in the history of the chemistry and geography of the Earth. As creative expressions of the implicate order, they *must* occur widely throughout the cosmos.

The notion that implicate patterns might find wide expression suggests the possibility of cosmic archetypes: patterns or forms that repeat throughout the explicate order. An example is the shape of a spiral, which recurs in sea shells, patterns of seeds in flowers, and spiral galaxies. In his latter years, Jung came to feel that archetypes are "metaphysical," transcending and incorporating both the mind and the physical cosmos. Barry McWaters tells the following story of a friend who, one night,

> had been watching a TV program about galaxies and noted that certain basic shapes recurred—spherical, spiral, elongated, etc. . . . The following night, the same channel . . . offered a program about micro-biological life in a swamp. Ecologists were researching the basic life support systems of the swamp, trying to identify the tiniest units in the chain of life. In describing their work, they noted that the organisms also took four or five basic shapes, which he noted, were almost identical to the galactic forms he had seen the night before.[57]

If there is a single conclusion to be drawn from all this, it is that the emergence of parallel themes, such as seen in many instances of synchronicity, is possible. In fact, it would be strange if a universe such as depicted by Bohm, Sheldrake, and Laszlo did not exhibit such parallel occurrences.

If we are indeed entering a vista—a mythos—of reality in which such ideas as morphic fields, psi-fields, and cosmic creativity have truth, then surely we have just started to explore the landscape. There must be many possibilities which we haven't begun to touch. While Sheldrake's concept of morphic fields emphasizes the static, stable, long-term aspect of patterns of form or of process, Bohm's concept of creativity emphasizes the opposite aspect—the tendency of the cosmos to produce new and original patterns, and on a nonlocal basis. These need not all be so profound as the creation of life itself, or even the universal expression of the spiral. Some are lyrical, as when Carl Jung encountered fish in numerous instances over a period of two days while he was working on a manuscript about the symbolic meaning of the fish symbol[58] (we will return to this story). Were his thoughts of fish and their symbolic meaning somehow enfolded into the implicate order to unfold again as various forms of fish, like the dream of the scarab unfolding again as a real beetle?

Mythologist Joseph Campbell recounted a related instance involving the praying mantis. He was at his home, a fourteenth-floor apartment in New York City, while reading about the mantis, which plays the part of the Hero in Bushman mythology. He sat near a rarely opened window that faced Sixth Avenue:

> I was reading about the praying mantis—the hero—and suddenly felt an impulse to open the window. . . . I opened the window and looked out to the right and there was a praying mantis walking up the building. He was there, right on the rim of my window! He was this big [showing the size]; he looked at me and his face looked just like a Bushman's face. This gave me the creeps![59]

Here we are coming very close to one of the fundamental enigmas of synchronicity: Meaningful coincidences frequently involve several events which, though radically different in form—one may be an idea, another a physical object—are tied together by a common pattern or theme, in the above examples the theme of fish and the theme of the mantis. On a still more profound level, such themes seem bound together by a common sense of meaning, the meaning of fish for Carl Jung as he worked on his manuscript and the meaning of the mantis for Joseph Campbell as he studied the Bushman mythology. Does the implicate order indeed frame abstractions and meaning as variously formed patterns which unfold into explicate reality?

As Bohm continues to develop his thinking regarding the implicate order, such a possibility becomes more credible. Recently he has discussed the idea of a superimplicate order which holds a position with respect to the implicate order much like the implicate order holds to the explicate. The superimplicate order acts on the implicate order as a kind of organizing principle, much like the thoughts of an author act on the ink and paper, organizing them into a specific written text. Bohm says, in fact, that the superimplicate order acts on the implicate order as consciousness acts on the brain. Consciousness supplies patterns of meaning and intelligence or, as Bohm says, *significance.* The superimplicate order is itself an expression of an even deeper super-superimplicate order, and on and on, as far as the mind can stretch. These deeper orders seem especially compatible with the idea that highly abstract and meaningful patterns—the patterns seen in synchronicity—may exfoliate into the explicate order in a wealth of forms.

These deeper orders portray a cosmic analogue to the human unconscious. In the human mind it is the unconscious that contains those powerful agents of potential for images or patterns, the Jungian archetypes. For instance, the Great Mother is one archetype that can express itself in many specific ways: as the goddess image depicted in archaic stone figurines; as an animal such as the bear, the rabbit and the cow; as places such as a plowed field or a deep well; as a flower such as the rose.[60] In the conscious mind the mythic image of the Great Mother takes on specific forms in specific situations. Something similar seems to be going on at the cosmic level, as in the recurring pattern of the spiral as galaxy, whirlpool, and seashell.

Bohm's vision of the holographic cosmos transcends the traditional distinction between mind and matter, consciousness and physical events. And so it tells us something of the origin of meaning and creativity in our own consciousness as well as in the objective universe without: they are the same. The deep well of the superimplicate breathes forth meaning which might rightly be termed mythic into the whole world of physical and mental reality. From this perspective there is no fundamental enigma in synchronicity, which so often juxtaposes meaning in the world of the mind with meaning in the world of objective events.

An Unidentified Flying Object

A decade or so ago in England, Suzanne Padfield[61] experienced a series of dreams in which she found herself lost in an unfamiliar countryside. While she was trying to find her way, a huge, bright, saucer-shaped craft would appear above her in the sky. She would then mysteriously find herself on a particular stretch of road near home. This recurring dream persisted for several months.

Driving home late one night, she attempted to take a shortcut and found herself lost in a maze of country lanes. Stopping at the top of a hill to read a signpost, she saw a huge, bright, saucer shape appear in the sky above her. Terrified, she drove toward the distant lights of the nearest town and, to her astonishment, found herself on the same stretch of road she had seen in her dreams! The next day the newspapers were full of reports of UFO sightings.

This phenomenon, in which a dream episode matches what appears to be a real external event, has obvious synchronistic overtones. Its most instructive feature, however, lies in Padfield's interpretation. Before explaining this, we should say a bit more about Padfield, as she has a very unusual qualification. She has a demonstrated talent for *psychokinesis*, the ability to mentally influence physical events such as the roll of dice or the motion of a suspended mobile. When she does this, she feels that she is, in fact, choosing one of several possible outcomes of which she is aware. "There is the subjective sense of exploring possibilities rather in the way one might remember what one did yesterday and the things one might have done in retrospect. I must emphasize here that I feel myself to be a part of these processes and not separate from them."[62]

From this sense of imaging and choosing a possible outcome in the psychokinetic situation, Padfield developed the idea that thought forms have a certain power to bring external realities into existence. This does not mean that everything we think about is likely to come true. As Padfield indicates, in the psychokinetic situation she seems to choose from a variety of possible outcomes rather than forcefully to assert some entirely new possibility. The effectiveness of the thought in bringing about an actuality increases, however, with the number of times it is thought, as well as with the number of people who think it. Suppose, for instance, you dream that your upcoming flight to Kansas City crashes. According to Padfield, this does not mean that the plane will actually crash, though the probability is increased by the fact of the dream. Suppose, however, that you cancel your reservation on account of the dream. This will further increase the chances of a crash, since in your thoughts you are focusing specifically on that possibility.

From Padfield's point of view, many cases of precognition, or knowing the future, are not precognition at all, but the very conditions that bring about the events in question. The same might be said for instances of synchronicity in which a thought is followed by the concrete appearance of the object of the thought. The appearance of the beetle at Jung's window was just such an instance.

Padfield relates a synchronistic event of this sort. "Once during an afternoon nap, which included a dozy dream about pigeons, a pigeon smashed through a plate glass window just feet from where I lay."[63] Dame Rebecca West reported a similar case to Koestler. While in the south of France she was writing a passage about a girl finding a hedgehog in her garden when she was interrupted by a servant who asked her to come see the hedgehog just found in the garden.[64]

Padfield's experience with the UFO seems exactly in line with the idea that thoughts can increase the probability of an actual corresponding event, no matter how unlikely the event may seem. Her repeated dreams and subsequent concerns regarding the appearance of a UFO over a particular stretch of country road apparently manifested themselves as something real enough to be observed by many others.

Her idea of synchronicity is consistent with Jung's notion of it as an outward manifestation of a process that also has an inner psychological component, though, as we will see, in Jung's case this always involved archetypes. (Jung, in fact, took considerable interest in the UFO phenomena, regarding them as projections of an archetypal image of the mandala, representing the completely developed personality.) Padfield's explanation of the UFO is consistent also with our earlier discussion of the possible role of the implicate order in synchronicity—the notion that a present thought may enfold into the implicate order and unfold later as an external reality. Perhaps the process can also work the other way. An external event may enfold into the implicate order to emerge again as a thought or idea. This could account for a common instance of synchronicity in which we find ourselves thinking of someone, perhaps an old friend, only to find that he has already mailed a letter to us or at that instant is calling us on the phone.

Such speculation leads directly to the prospect that virtually all ESP phenomena—telepathy, precognition, psychokinesis—can be regarded as special instances of synchronicity. Koestler, in *The Roots of Coincidence*, was the first to realize clearly that if one pushes the limits of synchronicity one eventually includes most of the traditional topics in parapsychology.[65] It may well be, in fact, that there is only one topic in parapsychology, and that topic is synchronicity. It is not our purpose in this book, however, to explore such a vast territory. It is enough to consider cases of synchronicity in which other causes, such as telepathy, are not the more obvious explanations. Perhaps when synchronicity itself is better understood its limits will be more easily defined.

Padfield believes that the influence of thought upon material events is actually based on physical brain processes—that patterns of brain activity occurring at the molecular or atomic level tend to bring about similar patterns of external world activity. Such patterns "are connected in a similarity space in which distances are defined by degree of similarity *and where time and space do not automatically appear at all*" (the emphasis is ours).[66] It sounds as if Suzanne Padfield would be quite at home in the company of Sheldrake, Laszlo, and Bohm. She could very well be talking about morphic fields, psi-fields, or quantum potentials.

Padfield says that her ideas were originally stated by the ancient yogic scholar Patanjali, who in about 800 B.C. wrote in the *Yoga Sutras*, "there is identity of relation between memory and effect-producing cause, even when separated by species, time and place."[67] Here, Padfield suggests that "species" is best translated as "identity of relationship." If her reading of Patanjali is correct, then we must admit that our current theories are something short of original.

Schrödinger's Cat

One interesting consequence of Padfield's observations is that various future possibilities can exist in a kind of limbo, exhibiting themselves as reality when there are sufficient matchups with the atomic or molecular patterns in human brains. When such patterns gain sufficient impetus, the corresponding possibilities become manifest reality. Padfield comments:

> One reads of biplanes being seen years before the first airplanes were invented. One reads of air balloons being seen years before the first airships, and now, machines that defy gravity, before what? I believe the thoughts each one of us considers private and our own are influencing, through their components, what we come to accept as the reality of the world we live in.[68]

The notion that possible futures can exist in a kind of suspended state is an old one in quantum physics. It is demonstrated in a famous thought experiment that was first suggested by quantum physicist Erwin Schrödinger. In this experiment, a cat—commonly referred to as "Schrödinger's cat"—is placed in a totally sealed box with no windows. The box also contains a device which, if triggered, will kill the cat. This could be, say, a canister of poison gas set to release on cue. A single particle, say an electron, is injected into the box through a tiny hole. Depending upon the action of the electron, a probabilistic affair, the device may or may not be triggered. The question is: *Prior to opening the box* (making an observation), is the cat dead or alive? The obvious answer is that, though we do not yet know, one or the other of the two possible outcomes is already true: the cat is alive or it is dead. An alternative interpretation, from the quantum perspective, is that the cat is *neither,* but that both possibilities exist side by side until the box is opened; then one or the other of them will become manifest.

An entirely different interpretation, offered by Hugh Everett and John Wheeler in 1957, is that at the moment the box is opened the universe splits in two, creating one universe in which the cat lives and one in which it is dead. This is the "many worlds interpretation." Its implication is that the cosmos is constantly dividing into an infinity of branches, each continuing to unfold its own history in parallel with the others. Something akin was described in 1937 by Olaf Stapledon in *Star Maker:*

In one inconceivably complex cosmos, whenever a creature was faced with several possible courses of action, it took them all, thereby creating many distinct temporal dimensions and distinct histories of the cosmos. Since in every evolutionary sequence of the cosmos there were very many creatures, and each was constantly faced with many possible courses, and the combinations of all their courses were innumerable, an infinity of distinct universes exfoliated from every moment of every temporal sequence in this cosmos.[69]

Padfield presents us with a vision of the future consistent with the idea that possible futures exist in a limbo state, like Schrödinger's cat, until just one of them is actually selected. The awesome alternative is that we may occasionally snatch glimpses of many possible futures, each to exfoliate in its own historical reality. Such ideas are even stranger than quantum theory, because at least in the latter the possibility of actually viewing potential or alternative realities is specifically forbidden.

Simultaneous Hypotheses

It is perhaps a synchronistic circumstance that, in 1981, three hypotheses, each postulating a tendency for patterns, once created, to reproduce themselves on a nonlocal basis, were published independently. Two of these we have seen: those of Sheldrake and Padfield. The third is that of Arthur Chester.[70] Chester's hypothesis was proposed to explain certain effects commonly observed in experiments in parapsychology. It also has direct application to synchronicity.

Like Padfield, Chester proposes that patterns of matter tend to reproduce themselves and that thoughts are such patterns represented in the physical matter of the brain. His approach differs from Padfield's in that he stresses the tendency of such patterns to assert themselves, not in external manifest realities but in the creation of similar brain states at other times and places. One interesting aspect of Chester's theory is that it includes not only the idea of nonlocality but also explicitly allows an arrow of causation to move either forward or backward in time. According to his view, present brain states may be influenced by future ones. This notion does not seem entirely incompatible with Padfield's idea that possible futures can exist in the present in a kind of suspended state, perhaps even influencing the mind in the present. It is, however, quite contrary to Sheldrake's thesis that morphic fields develop with time, a thesis specifically designed to provide testability to the hypothesis of formative causation by conventional scientific experimentation. Chester's ideas, on the other hand, were designed especially to deal with parapsychological effects in which time seems to work in unusual ways. For instance, a subject in a card-guessing experiment might call out the sequence of playing cards pulled at random from a deck with an accuracy well beyond what would be expected by chance alone. The cards, however, may not actually be pulled until the following day. This disregard for the normal temporal order of things is all too common in extrasensory perception experiments.

Chester and Padfield both rely on the notion that particular mental states, especially mental images, can be equated to configurations of events in the physical brain. They take for granted, however, a perfect or nearly perfect correspondence between mental states and brain states. Their assumption is warranted only if we assume that the brain and mind are identical. If we take any other point of view, say, the dualist view that the brain and mind are separate but interacting processes, then it is not obvious at all that a particular mental state is always associated with a particular brain state.

It seems to us that the conscious mental state itself "looks" a lot more like the physical event than any brain state does. For instance, the dream image of the golden scarab beetle in the mind of Carl Jung's patient had more in common, structurally, with the real beetle than did the corresponding combination of electrical and chemical processes in her brain. If we allow mental events some degree of reality, why appeal to brain states which may or may not be associated with them? Bohm seems comfortable with this. In his theory, consciousness and mental life in general are equal among the creative products of the implicate order.

LEVELS WITHIN LEVELS

The Omega Point

The experience of synchronicity gives us the sense of being on the receiving end or downstream side of a much greater process, a process that has not only breadth but also a vertical, or value, dimension. The breadth is the size of the process in terms of the numbers of persons and events it touches. The dimension of value is experienced subjectively as a sense of meaning, purpose, or wonder—in Jung's terms, *numinosity*.

The idea that the cosmos has a vertical structure as well as a lateral one is historically very old. Aside from visions of higher realms such as the dwelling place of the gods or departed souls, in certain instances such vertical structures interact directly with the physical world of the living. We have already seen, for example, Plato's higher realm of perfect forms, operating in a fashion somewhat similar to Sheldrake's morphic fields, to give form to the less-than-perfect objects of this world. Moreover, it defines the essence of a thing. A horse, for instance, can easily be recognized as a horse and not, say, a huge dog or humpless camel, not because all horses are alike, but because all horses take their basic form from the perfect horse of the higher realm.

We have previously touched upon the medieval notion of sympathies between the higher spheres and the earthly world of human affairs. These gave a sense of connection with higher dimensions of the cosmos entirely unknown to modern man.

Perhaps the most ambitious effort in recent times to visualize a world structure with a clearly vertical dimension, and one that has meaning for synchronicity, was in the work of Pierre Teilhard de Chardin, a Roman Catholic priest of the Jesuit order. Teilhard de Chardin first achieved prominence as a physical anthropologist for his contributions to the understanding of human evolution. In his remarkably brilliant mind, the budding of spiritual, and indeed mystical, consciousness in the presence of the rational search for the evolutionary origins of mankind resulted in a most dramatic vision of our place in the natural cosmos.[71]

Teilhard de Chardin did much of his work in the 1930s and 1940s. Science was still very much under the sway of the old mechanistic worldview, despite the fact that quantum physics had been postulated several decades earlier. It is a remarkable credit to his genius that he was able to stay largely within this worldview and nevertheless create a truly humanistic and indeed transcendent image of humankind's place in the cosmic order.

Something needed to be added to the mechanistic model to allow the unfolding of a vertical dimension in nature. This something Teilhard de Chardin termed "radial energy," a type of energy inherent in matter which "draws it towards ever greater complexity and centricity—in other words forwards."[72] Teilhard de Chardin was proposing a form of energy which draws matter into ever-higher levels of organization and complexity. A theory of this nature was revolutionary because the energy physics of the mechanistic worldview tended to place great stress upon the second law of thermodynamics, the law of entropy, which specifies that physical systems tend toward increasing disorder as time passes.

Actually, this law strictly applies only to closed systems, those that do not exchange energy with the environment, as living organisms do. Nonetheless, the second law has maintained great influence on scientific thought where the evolution of complex systems is concerned. Teilhard de Chardin's radial energy was intended to provide a vertical force against the downward drag of entropy. He realized, as did Sheldrake, that a very small amount of energy can go a long way toward the creation of form and, in fact, suggested that the impact of this radial energy is felt as a patterning influence, asserting itself by guiding the arrangement of things, "for a highly perfected arrangement may require only an extremely small amount of work."[73]

Teilhard de Chardin believed that the cosmos presents us with two faces: an exterior, material reality and an interior reality, or consciousness. He wrote that "the exterior world must inevitably be lined at every point with an interior one." This he thought to be true at all levels of material existence, from the single atom through complex chemical structures and on up through simple living organisms to highly complex ones, leading in a direct line to humankind. A progression is established in the exterior world leading from the simple to the vastly complex, and simultaneously, in the interior world of consciousness, from the separate and elemental to that which is large and rich in quality. "Whatever instance we may think of, we may be sure that. . . . A richer and better organized [physical] structure will correspond to the more developed consciousness."[74] Such a richer and better organized structure is the human nervous system.

The thing that is exciting about Teilhard de Chardin's work is not so much his marriage of matter with consciousness—which means that the evolution of complexity in the physical world heralds the evolution of quality in conscious experience—though this idea alone has been profoundly influential in the scientific intellectual community. (Witness Arthur C. Clark's attribution of consciousness to Hal, the supercomputer in 2001: A Space Odyssey, largely on the basis of the computer's awesome complexity.) The exciting thing is Teilhard de Chardin's projection of evolution well beyond humanity in its present condition; he, in fact, speculated about the future evolution of the Earth itself, including all its life forms, and especially humankind.

To Teilhard de Chardin, the total mental activity of humanity may be thought of as a web, or perhaps a membrane, that encircles the entire earth. This noosphere, or sphere of mind, is analogous to the biosphere, the sphere of organic life on Earth. The noosphere is the inner side of nature, the side of consciousness and mind. As human societies develop, there is an increasing global centralization of this internal dimension, the realm of mind. Individuals come closer together in their activities, interactions, and communications. This process, which today is so apparent in our external global culture, has an internal aspect, and this internal coming together of minds on a planetary scale gives birth, in Teilhard de Chardin's vision, to a new and higher level of being, a global consciousness, the *omega point*.

The omega point represents a quantum leap in global evolution. The omega point is to the individual minds which form it as the individual human mind is to the neurons which form the brain. Like the human mind, the omega point has its own emergent and qualitatively higher properties. It has a transcendent mystical dimension, and at the same time it unifies and centralizes the activities of its constituent minds in a fashion not unlike that in which the activity of the individual human mind draws together and centralizes the activities of the nerve cells of the brain. This process occurs, not through loss of individuality but through a mutual enfolding of the most personal inwardness of each individual with other individuals. In a word, the omega point is the fruit of that most essential of inner experiences—love.

It was Teilhard de Chardin's belief that the omega point is not something which might possibly come into being in some hopeful future; its creation is taking place at this very instant of evolutionary time, and its deep personal and mystical dimensions tend to draw us toward it. Its organizing influence is already felt as a presence in the world.

It does not require great imagination to wonder just how the omega point can serve as a focus of organization and coordination of processes, both physical and mental, throughout its global presence. Some further principle of organization seems needed.

Ira Progoff, one of Carl Jung's most accomplished students of synchronicity, suggests that something is missing in the way Teilhard de Chardin conceptualizes the organization of noosphere activity with regard to higher levels of evolution. In the first chapter of *Jung, Synchronicity, and Human Destiny*, Progoff suggests that synchronicity fills the bill very well.[75]

Here we have a perspective on synchronicity as a direct manifestation of a higher reality, the organizing presence of the omega point.

It represents a bridge to a higher, more luminous level of being. In later chapters we will see that Jung, in his own way, also saw synchronicity as the manifestation of a higher principle.

SELF-ORGANIZING SYSTEMS

In 1977 the Nobel Prize in chemistry was awarded to Ilya Prigogine for his contributions to the understanding of chemical systems that are not in equilibrium. More specifically, Prigogine has shown that complex systems (chemical and otherwise) when placed in a flow of energy, can restructure themselves into higher orders of organization. A simple example is a pan of water that is slowly heated on the stove. Chemically, the water is composed of individual molecules bouncing about. This bouncing about (termed Brownian movement) becomes more vigorous as energy is introduced, the water heats up, and at a certain point an entirely new level of organized activity is seen. This is the appearance of a complex pattern of convection currents which form a new and higher degree of organization. The individual molecules still bouncing around as individuals are carried about in the large currents, becoming part of them without losing their own identities. This tendency for higher-order structures to emerge from lower-order ones is seen in the behavior of complex patterns of individual cells within an organism, and even in the behavior of entire societies of organisms.

For example, if individual heart cells are placed in a petri dish, each contracts with its own individual rhythm. When some critical number of cells is reached, however, all fall into a synchrony which is characteristic of the normal action of the heart. Likewise, a few ants placed in a sandpile wander about aimlessly, apparently oblivious to each other. Continue to add new ants, however, and at a certain point they begin to self-organize into a working society, each assuming its particular role in the larger structure. These ideas are discussed in detail by Erich Jantsch in *The Self-Organizing Universe*.[76]

Prigogine's work, and that of others such as Erich Jantsch,

embodies Teilhard de Chardin's earlier ground-breaking thought. Prigogine is above all a mathematician, and so his formulations fit the form most desired by modern science. Thus, while Teilhard de Chardin was influential in the private thoughts of many scientists and philosophers, Prigogine's work has quickly attracted attention in the arena of scientific discourse. It is not apparent that the ultimate fruit of Prigogine's labors will have the transcendent qualities of the omega point. Indeed, the highest levels of organization in Prigogine's universe remain to be explored.

What is clear from Prigogine's work, however, is that the cosmos must be viewed as many-tiered, with vast planes organized vertically in a grand array. Each level, the life of individual cells of the human body, for example, is a self-contained existence. Such cells, however, form the building blocks of a community of organs—the heart, the liver, and so on. Each of these has a life of its own, while as a group they form the basis of a greater whole, the human body. The result of this total structure is an entirely new emergent property called mind. In the outer world, individual minds form societies of persons. According to Teilhard de Chardin, in the inner world they form the noosphere which gives birth to the omega point.

The notion that processes in nature are best understood as many-tiered, or hierarchical, was actually introduced during the first half of this century, long before Prigogine's work was published. The person responsible was an influential yet surprisingly little-known individual named Ludwig von Bertalanffy, regarded as the father of general systems theory.

We have, in fact, already covered much of the ground stipulated by general systems theory, taken in its broadest form. The basic insight is that complex processes (organisms, societies, economies) are comprised of many levels, each of which functions according to its own internal laws and restrictions while participating in the greater whole. The heart beats according to its own internal regulating processes while participating in the greater human organism which, as a unit on its own level, has its unique existence with its own behaviors. All of this is familiar to us. What Prigogine added to von Bertalanffy's vision was the idea of a thrust in the very essence of matter to evolve new and higher levels of organization and unity.

The possibility that synchronicity may be an emergent property of a higher level of physical or mental organization is exactly what Ira Progoff was suggesting when he postulated that a new, emergent principle of organization was needed in the omega point. Systems theorist Erich Jantsch actually suggested quite independently of Progoff that synchronicity may be a principle that evolves out of the highest levels of cosmic organization. It was Arthur Koestler, however, in his lifelong passion for understanding synchronicity, who most clearly postulated synchronicity as the fruit and substance of a high emergent level of unity in human affairs.

In his own colorful and creative manner, though deeply influenced by von Bertalanffy, Koestler built a unique conceptual system around a unit he called the "holon" (from *holos*, "whole").[77] Koestler conceived of the holon as both a subunit in a greater system and a complete entity in and of itself. The holon exhibits the quality of autonomy while at the same time functioning as part of a larger whole. Now all this seems quite familiar, but the unique emphasis in the concept of the holon is its self-assertive tendency:

> The living organism and the body social are not assemblies of elementary bits; they are multi-levelled, hierarchically organized systems of sub-wholes containing sub-wholes of lower order, like Chinese boxes. These sub-wholes—or "holons," as I have proposed to call them—are Janus-faced entities which display both the independent properties of wholes and the dependent properties of parts.[78]

Koestler believed that synchronicity springs from the highest integrative potential on the human level. Thus "by regarding the integrative tendency as a universal principle which includes acausal phenomena, the picture becomes greatly simplified, even if still beyond the grasp of understanding."[79]

PLUMBING THE HOLOMOVEMENT

Bohm states that the sequence of orders—explicate order, implicate order, superimplicate order, and even deeper orders—do not represent a hierarchy. In one sense, we can understand what he means. It would be entirely wrong to imagine these various orders in any literal way as planes of reality, stacked one on top of the other. Bohm wants to be very clear that each level totally penetrates the other levels: there is only one reality.

But in another, compelling sense, Bohm's model, in fact, does describe a hierarchical structure. The explicate order that we experience daily constantly unfolds from the implicate order. This, in turn, unfolds in a similar flowing motion from the superimplicate order, and so on. Each level of this manifold fabric of orders is simply another aspect of reality.

More than one author has speculated that during moments of unusual penetration, we may actually look into the implicate process, sensing information that is not available in the explicate order. Psychologist and futurist David Loye refers to this as the ability of the mind to "range" in the implicate order.[80] At these times the conscious mind taps into the timeless well of the cosmic fabric. We might imagine this as the basis for certain extrasensory experiences that are not bound by location in space and time. The most reliable form of ESP, for instance, is "remote viewing," in which a person in the laboratory is asked to imagine what someone else is seeing.[81] The other person can be anywhere in the world. Is this done by ranging in the implicate order?

A less dramatic but more important process that may involve ranging in the implicate order is intuition. As Goldberg points out in *The Intuitive Edge*, certain instances of intuition seem to penetrate beneath the visible reality of the explicate order.[82] Intuition can involve a sense of contact with a more surefooted but ironically ineffable reality from which one perceives and acts with a precision and confidence completely unwarranted by the explicit facts of the situation.

For instance, John Walton, who hunted man-eating tigers for the state of Bengal in India during the first years of this century, discovered more than once as he tracked through the jungles that he had become the prey himself, and upon the prompting of some sixth sense turned around just in time to save himself by avoiding being attacked from behind.

Ability to range in the implicate order would in no way require the mind to go outside itself in search of data which might be widely scattered. As we have seen, the holographic model suggests that the entire cosmos is enfolded into itself in each part. The founding text of Ch'an (Zen) Buddhism, the *Platform Text*, expresses this idea beautifully:

> Fundamentally, there is only one Great Ultimate, yet each of the myriad things has been endowed with it and each in itself possesses the Great Ultimate in its entirety. This is similar to the fact that there is only one moon in the sky, but when its light is scattered upon rivers and lakes, it can be seen everywhere. It cannot be said that the moon has been split. [83]

One need only have eyes that can look within.

It is evident that our understanding of the cause of synchronicity and its meaning in our lives is limited not only by our understanding of physics but also by our understanding of the mind, and indeed even our understanding of the brain. We must seek a final under-standing—at least for this book—only after looking at that most enig-matic entity, the human brain, and seeking the knowledge that psychology has to offer us.

SILENT RESONANCE:
the mind and the brain

*The marvelous and mysterious which is pe-
culiar to night may also appear . . . in the
remarkable silence that may intervene in the
midst of the liveliest conversations; it was said,
at such times, that Hermes had entered the
room. . . .*

WALTER OTTO
The Homeric Gods

THE ROOTS OF LEFT AND RIGHT

The past few decades have marked dramatic advances in our understanding of that most remarkable structure in the known universe, the human brain. It is natural that we should wish to make use of this burgeoning knowledge in developing our understanding of synchronicity. At a minimum, we would like to appreciate something of how brain processes are related to synchronicity. At best, we might find clues to the roots of synchronicity itself.

There seems to be a subtle though persistent desire of nature to express herself in slightly asymmetrical forms. Though the right and left hands appear to be mirror images of each other, they are not, nor are the feet. As most of us know from trying on shoes, one foot is larger than the other. In the last century, Louis Pasteur noted that while tartaric acid molecules come in two shapes, one the mirror image of the other, a certain plant mold he was studying acted on only one of the forms. He concluded, "This important criterion [asymmetry] constitutes perhaps the only sharply defined difference which can be drawn at the present time between the chemistry of dead or living matter."[84]

Indeed, the importance of asymmetry in living matter is attested to by the fact that the molecules most basic to life, the protein molecule and the double helix, are asymmetrical, winding in only one direction. In the past few decades it has become increasingly apparent, both in the laboratory and in theory, that this absence of perfect symmetry in nature runs deep, cutting all the way to the form of the atom itself. It should be no surprise, then, that asymmetry is at evidence in the human brain, or that such asymmetry might reflect a complementary asymmetry in human psychology, giving each of us two separate, complementary mental qualities.

In 1861, the French neurologist Paul Broca presented to the Society of Anthropology in Paris the brain of a man who had, prior to his death, suffered a gradual loss of speaking ability. The brain exhibited a damaged area in its left frontal region. "Broca's brain," has become the most famous brain in history because it marks the beginning of our understanding that the human brain is asymmetrical in its functions.

Broca observed that in most persons the left side of the brain controls language. Language is not only vital to human communication, but it is also the very stuff of logical thought. Without it there would be no science, no mathematics, no books of any kind; very little could be passed from generation to generation, and there would be no human culture as we know it. Broca's work indicated for the first time that all this springs from the left side of the brain—the left hemisphere.

What about the right hemisphere? As early as 1865, the British physician Hughlings Jackson, a giant of nineteenth-century clinical neurology, was to write that if it "be proven by wider experience that the faculty of expression [language] resides in one hemisphere [the left], there is no absurdity in raising the question as to whether perception—its corresponding opposite—may be seated in the other."[85] If the word "perception" is taken in the larger sense of a holistic grasp or understanding of a situation, then Jackson was quite correct. As we know today, the right hemisphere is vitally involved in holistic or global thinking.

Current scientific knowledge concerning the right and left hemispheres suggests a broad picture in which the left brain is the logical, analytic, rational thinker. The right brain seems to be the holistic, perceptual, creative thinker; it is not constrained by the dictates of linear logic but is able to grasp and act on situations as wholes. The right side is vital to art, creativity, and emotion. All this applies to the majority of persons, those with left hemisphere language (95 percent or more of right-handers and 70 percent or more of left-handers). For others, the situation is reversed or the hemispheres are mixed in function. While these distinctions are a bit overdrawn, they provide us a starting point for exploring theories of synchronicity that are based on the two-chambered or "bicameral" structure of the brain.

That the two sides of the brain are distinct in function has been known for hardly over one hundred years. Such factual knowledge, however, echoes a much older intuitive sense for differences between the meanings of the words "left" and "right." To understand this, we need to note that the brain is connected or "wired" to the body so that the left hemisphere controls the right side of the body and the right hemisphere controls the left side. The left brain, for example, controls the right hand, while the right brain controls the left hand. This is why a severe stroke to the left (language) brain can affect not only an individual's speech but also paralyze the *right* side of the body. Along with this crossover of control there is a similar crossover of sensation, so that what is felt on the left side of the body is registered by the right side of the brain, and vice versa. With this as the key, we can understand certain differences between the symbolism of right and left that have been known intuitively for many years.

Psychologist Jerome Bruner, a pioneer in the understanding of creative thought and intuition, wrote in 1962:

Since childhood, I have been enchanted by the fact and the symbolism of the right and the left—the one the doer, the other the dreamer, the right is order and lawfulness, *le droit*. Its beauties are those of geometry and taut implication. Reaching for knowledge with the right hand is science. Yet to say that much of science is to overlook one of its excitements, for the greatest hypotheses of science are gifts carried in the left. [86]

David Loye, among others, has summarized many of the qualities traditionally associated with the left and the right. [87] Those often associated with the right side (left brain) are verbal, analytic, and sequential knowledge: that which is intellectual, propositional, and focal. The right is also associated with that which is known, explicit, out in the open, and with the male principle and the sun. On the other hand, those qualities often associated with the left side (right brain) include spatial/visual, Gestalt/holistic, and simultaneous knowledge; that which is sensuous (directly sensed); appositional (all elements existing together at once); and diffuse. The left is also associated with the implicit, unknown, mysterious, fertile, creative, and with the female principle and the moon.

THE TOMB OF THE GODS

A creative and articulate theory of synchronicity based on brain function was offered by Barbara Honegger in a paper presented at the 1979 meeting of the Parapsychology Association. In it, she proposed that meaningful coincidences are externalized, dreamlike processes controlled by a second language center located in the *right* hemisphere of the brain. [88]

The notion that the right hemisphere may have a language center of its own, separate from that of the left hemisphere was not new. In 1976, psychologist Julian Jaynes published *The Origin of Consciousness in the Breakdown of the Bicameral Mind*. In this book he argues that in ancient times, prior to about the first millennium B.C., the right hemisphere of the brain had a language center which functioned quite independently of the left. This center spoke to the left brain in powerful and demanding voices via certain of the large neural pathways that connect the two hemispheres. The left hemisphere, which Jaynes identifies with the conscious self, heard the voices of the right hemisphere as coming from somewhere outside itself— perhaps from a cave, from the sky, or most likely from the statue of a god. These were the voices of the gods. [89]

Such voices, according to Jaynes, spoke with absolute authority to the conscious self of the left hemisphere, giving instructions regarding virtually all major decisions that had to be made. The records left by the ancient Mesopotamians are filled with specific references to such directives. For instance, an old Sumarian proverb says, "Act promptly, make your god happy." (Keep in mind that in these ancient civilizations each individual was accustomed to answering to his or her own personal god, usually kept somewhere about the house in the form of a statue. The average person had little intercourse with the great gods.) That these communications were real to the persons who experienced them is evidenced in many of the cuneiform texts. From about 1700 B.C., for example, we have an account of the goddess Ningal as "counsellor, exceeding wise commander, princess of all the great gods, exalted speaker, whose utterance is unrivaled."

Jaynes felt that decision making is ideally suited to the right brain, with its ability to grasp whole situations, that is, to take all factors into account at once. He felt that the left brain was better suited to carrying out these decisions, working out the details as it went along. In the ancient agricultural civilizations of the Middle East, many such decisions had to do with the more-or-less predictable demands of farming. During the second millennium before Christ, however, this situation changed. This dark millennium brought for the first time in history large-scale and widespread warfare between nations. The directives of the gods dealt increasingly with the topic of war. It was also a millennium of frequent natural disasters in the Middle East. The resulting widespread disruption of the previous social fabric taxed individual survival skills to their maximum. During this chaotic period of history, a unified mind, capable of greater flexibility, was a great advantage. Such a mind developed at the expense of the voices of the gods. The brain became whole through greatly increasing communication between the two hemispheres, and the voices gradually disappeared. Today, Jaynes speculated, only a small residual part of the original right-brain language capacity remains.

Much of the evidence for Jaynes's theory comes from the work of the Canadian neurosurgeon Wilder Penfield.[90] In the 1950s, Penfield observed that during brain surgery mild electrical stimulation to the right-brain area that corresponds to the left-brain language region produces hallucinated voices. Though Penfield believed that he was activating memories locked in the brain, often since childhood, Jaynes presents evidence to the contrary. For instance, while in some cases the voices seemed familiar to the patient, more often they did not. In the majority of instances the patient did not recognize the voices, which were usually hazy and indistinct. Jaynes believed that in stimulating the linguistic region of the right hemisphere Penfield had reactivated the ancient mechanism for the voices of the gods. He had, in a sense, entered the tomb of the gods and was listening to their ghosts.

Today these voices are normally silent. Jaynes believes, however, that they are still heard in the hallucinations of schizophrenia, in whose many sufferers the old mechanism has somehow come to life. He notes that prior to the advent of modern drug therapy the majority of those with schizophrenia experienced such voices. They tended to carry great authority for the schizophrenia victim who, with only the greatest of efforts, could willfully ignore them. Moreover, the voices were demanding, as the gods had once been. And like the gods, they often seemed to come from outside the individual, from somewhere else in the room, or in some instances even lacked spatial location entirely.

Like Jaynes, Barbara Honegger believes that there is a right-hemisphere language center that is alive and well. She feels that it asserts itself as the author of night dreams and also as the architect of meaningful coincidences. Now, the notion of a right-hemisphere language center is highly speculative, despite Penfield's observations during surgery. Honegger's theory does not actually require that the right hemisphere have a fully developed speech facility. Rather, it must have at a minimum its own "deep-structure" linguistic process. Linguists use this term to refer to the meanings and relationships expressed by a sentence, as opposed to the literal form of the sentence as it is spoken or written (the surface structure). A sentence in French, for instance, may have the same deep structure as an equivalent one in English, while their surface structures are quite different. Honegger views right-brain hemisphere deep structure as the source of both night dreams and synchronistic coincidences as well as of a variety of types of paranormal phenomena.

Interestingly, Honegger's theory does not actually take its origin from Jaynes's, or even from Penfield's work, but rather from an examination of the content of dreams as compared to the content of synchronistic coincidences. She points out that since the time of the publication in the late nineteenth century of Freud's classic, *The Interpretation of Dreams*, certain major theories of dream interpretation have emphasized the idea that dreams hide a deep structure of a basically linguistic nature. In other words, the real meaning of a dream is to be found in a verbal formula that underlies it. A straightforward instance of this, and one that Freud believed to be common, is the depiction of verbal puns in dream images. In fact, one of the authors of this book had such a dream two nights before writing this passage. He went to bed quite late, completely exhausted, and slept deeply. The last thing he did before falling asleep was to watch an old *Twilight Zone* rerun about three astronauts who landed on an asteroid that turned out to be a cemetery maintained by some advanced civilization. No doubt this influenced the content of the subsequent dream. In it, the author found himself actually sleeping in a cemetery. The deep-structure message of the dream was that he was "sleeping the sleep of the dead." A referral to the dictionary disclosed that the word "cemetery" comes from the Greek, meaning sleeping chamber!

Honegger believes that synchronistic events can be interpreted in a fashion similar to dream interpretation, that is, by searching for their linguistic deep structures. An example of a synchronistic occurrence which involved a pun was reported a few years ago in Texas. It concerned a state highway patrolman who was severely injured in a motorcycle accident and the businessman who saved his life. The motorcycle patrolman, Allen Falby, had struck the back of a truck and was thrown to the road when the businessman, Alfred Smith, happened by—stopped his car to see if he could be of assistance. Smith found Falby bleeding severely from one leg. Using his tie as a tourniquet, Smith stopped the bleeding, thus saving Falby's life. It was five years before the two men met again, and this time it was Smith who was badly injured. He had been in a car accident. Falby was the first to arrive and, as fate would have it, he found Smith bleeding badly from one leg. Only after Falby had applied a tourniquet to stop the bleeding did he recognize the businessman. Later, he was heard to comment jokingly, "One good tourniquet deserves another."[91]

This example shows the way Honegger interprets synchronicity. In the Freudian tradition, she emphasizes hidden linguistic relationships, in this case a pun. In another example, Honegger also expresses hidden verbal meanings. It concerns an acquaintance of hers, a lady who found, to her surprise, that the word "congeniality" was not listed in her Webster's dictionary. She commented about this to her sister. Both were later removing books from the library shelves of their family home in preparation for selling the house when the lady remarked that she had just removed the novel Don Quixote. On the other side of the room, her sister noticed that the book was in her own hands at that very moment, and they both commented on the coincidence. Later, while sitting in a rocking chair in her father's room, the lady noticed a hanger on the floor and, bending over to pick it up, wrenched her back. Now, it turns out that their father was a kindly and dignified Don Quixote–type of gentleman of the family name Webster. He was prone to martyrdom in family affairs but was less than happy about the forthcoming sale of the house. Honegger comments:

> From these events I was able to predict that [the lady] felt deep frustration at a lack of congeniality in her family, and that both sisters felt much of the problem was due to their father who, like Quixote, felt called upon to redress the wrongs of the world. From the hanger, I made the prediction that a hanging had figured somewhere in the family history, which the father verified. His sister had hanged herself in spite of years of sacrificing his own family's interests in her behalf.[92]

Honegger reported that great energy was released in discussing these events.

Honegger's cases are impressive in their complexity. The instance just described is actually one of the simplest. While providing a fascinating approach to the whole topic of synchronicity, however, her ideas present some difficulties. To begin with, recent research does not substantially support the notion of a right-hemisphere language center. The electrically evoked auditory hallucinations of Penfield's patients, for example, can be reliably produced only in epileptics. The reason for this is unknown, but it casts doubt on the notion that the right hemisphere is a source of linguistic effects, at least in most persons.

Honegger, of course, does not claim that the right hemisphere can produce speech, only that it controls the deep-structure linguistic processes behind dreaming and synchronicity. Still, without Penfield's voices, the case for right-hemisphere language, beyond the known ability of that side of the brain to understand simple spoken sentences, is weakened. Moreover, recent research does not support the older suggestion that the right hemisphere is solely in charge of dreaming. And, when all is said and done, we are still left with the unsettled question of whether synchronistic coincidences can legitimately be interpreted as dreams are.

The most awesome aspect of Honegger's approach is the idea that the right brain can orchestrate events in the real world. Honegger comments that most synchronistic events are directed toward wish fulfillment. If this is the case, how far can these effects go? For instance, the businessman's car accident, which occurred five years after he saved the life of the highway patrolman is explained as having satisfied the patrolman's longstanding wish to repay his debt to his previous benefactor. (It would have been better to have sent a card!) We are reminded of the overwhelming possibilities of the dream-cum-real world of Ursula Le Guin's *The Lathe of Heaven*, a novel about a young man whose dreams immediately become reality, clearly in the interest of his own wish fulfillment. A related, but more humorous version of this theme was developed by Larry Niven in his novel *Ringworld*, which presents a young woman with a genetically acquired capacity for good luck. Unfortunately, others around her often have very bad luck because things tend to work out in the best and most straightforward way possible for meeting her own needs, with complete disregard for others.

Despite the prospect of unpleasant coincidences, it would be a serious mistake to turn our backs on the notion that synchronicity might act as an agent of wish fulfillment. The more fundamental notion that Honegger presents us with, however, is that the basic template for synchronicity is taken from the structure of the brain itself. This is a fascinating idea which, at this point in the history of brain science, we are simply unable to assess. Interestingly, it is an idea to which Jung himself would not have been opposed. In 1929, in his "Commentary on The Secret of the Golden Flower," Jung wrote that "the collective unconscious is simply the psychic expression of the identity of brain structure irrespective of all racial differences."[93] Jung further maintained that this expression explains the high degree of similarity found in mythic themes throughout humanity. Recall that such themes are based upon psychological archetypes which, in their turn, Jung believed to be the sources of synchronicity.

SILENT RESONANCE

Another avenue to understanding the connection between synchronicity and the brain involves the meditative state. There appears to be a link between synchronicity and meditation: as the meditative state deepens, synchronistic coincidences occur more frequently. In *The Global Brain*, Peter Russell observes:

> Many people who practice meditation of one kind or another have found that the deeper and clearer their meditations, the more they experience curious patterns of coincidences. This tends to be particularly so after extended meditation retreats; on returning to regular activity, each day can seem like a continual train of the most unlikely, and most supportive, coincidences.[94]

Evidently, prayer can have a similar effect. The British archbishop William Temple once commented, "When I pray, coincidences start to happen. When I don't pray, they don't happen."[95]

Some years ago it was popular to assume that meditation is a right-hemisphere process. Meditation certainly is not verbal beyond some minimal activities such as counting breaths or repeating a mantra and is associated with a holistic experience of oneself and the world and thus opposite to the logic and analysis usually associated with the left hemisphere. All this not withstanding, we believe that the state of meditation is, above all, a neurologically balanced one. As such, it is difficult to imagine that it would involve one hemisphere more than the other. Both hemispheres should be functioning in a relaxed but optimal state.

More recent research strongly supports our notion of a balanced brain state during meditation. Especially notable is the work of a group at the Maharishi European Research University headed by David Orme-Johnson.[96] Using electroencephalographs, this group measured brain-wave activity from several cortical locations during various stages of sleep, wakefulness, and transcendental meditation. They found that during meditation brain-wave activity is balanced between the two hemispheres. Not only do the EEG patterns from the two hemispheres become similar with regard to frequency, but the waves themselves become highly correlated. Put simply, the EEG activity of each hemisphere mirrors that of the other. This characteristic, termed *coherence*, seems to be a significant marker of the meditative state, and to our knowledge is found in no other state of mind. Orme-Johnson and his group feel that this brain state corresponds to *pure consciousness*, an advanced state of meditative awareness. While other types of meditation have not been studied as thoroughly, the general evidence strongly supports the idea that a balanced or synchronized brain state is characteristic of virtually all of them.

Pulling all this together, we have substantial reason to suspect that the mode of brain activity most favorable to synchronicity is the balanced, profoundly silent state experienced in deep meditation and prayer, a state which is accompanied by resonant coherence of the EEG rhythm. This is the silent resonance of the brain.

The Holographic Brain

The presence of coherent brain-wave activity during meditation suggests a holographic process. Creating a hologram requires the use of a special type of light, termed *coherent* light. Typically supplied by a laser, coherent light is characterized by the synchronous action of all light waves. Each is aligned with the others much as EEG waves are aligned during meditation.

For two decades the neurologist Karl Pribram has been arguing for the idea that certain higher brain processes depend upon a holographic mechanism.[97] He does not claim to know specifically which aspect of brain activity corresponds to the coherent laser light of a visual hologram. It is quite possible that such coherence arises in the complex interplay of the minute electrical ripples that ceaselessly dance across the surface of brain cells. Such activity may be mirrored in the EEG itself, especially during meditation. Whatever the source, Pribram believes that both perception and memory are based upon a holographic brain mechanism.

Several lines of evidence lend support to Pribram's view. First, perception seems subjectively much like a hologram. The experience of the visual world is one of depth and richness such as might be expected in a hologram, rather than, say, in a flat, two-dimensional photographic image. Moreover, current studies of perception and artificial intelligence emphasize the holistic aspect of vision. Engineers working on visual systems in robots are finding the sequential, item-by-item analysis typical of conventional computers unable to cope with the immense complexities of real-world visual events. They are resorting to holographic images to provide flexible, practical robotic visual systems.[98] There is also increasing evidence from neuroscience laboratories that the brain acts upon perceptual information in exactly the fashion one would expect if holographic processes are occurring. Specifically, holographic processes require that information—visual images for example—be transformed into a form that expresses frequency, or the rate of change in time or space. Recent studies of the visual centers of the brain show unmistakably that cells in these areas do, indeed, perform just such translations.

Other evidence comes from the study of how the brain stores memories. A consistent observation is that, within broad limits, damage to specific areas of the brain does not eradicate specific memories. After a stroke, for example, the victim does not lose memories of one specific individual and no one else. That is what might be expected if such memories were stored in individual locations, like images on a photographic plate. In fact, if brain damage is extensive, memory typically becomes fuzzy. This is exactly what one would expect if the brain operates like a holographic plate rather than a photographic one.

Recall that each part of a holographic plate contains its whole image. Unlike a photograph, when part of the hologram is damaged, the entire picture is still preserved in what remains. If a large portion of the holographic plate is lost, the image, like memory after extensive brain damage, becomes fuzzy. Holographic plates, moreover, are able simultaneously to store large numbers of images without mutual interference. This quality separates holographic plates from photographic negatives, on which multiple images result in a confused blur. High-capacity storage is another aspect of holography that makes it seem similar to human memory, which seems to be virtually limitless.

Let us suppose that under certain conditions the human brain has the capability, as suggested by David Loye in Chapter 2, to "range" in a much larger holographic reality, Bohm's implicate order.[99] It is reasonable to suspect that the coherent EEG pattern seen in deep meditation signifies a state of holographic brain activity which is conducive to such ranging. If, as was also suggested in Chapter 2, the implicate order contains universal patterns or cosmic archetypes such as the spiral seen in such diverse locations as sea shells, whirlpools, the arrangement of the seeds in sunflowers, and in spiral galaxies, then deep meditation is an optimal state for opening receptively to the presence of such archetypes. Perhaps the implicate order is also the receptacle of the blueprints for the psychological archetypes described by Jung and encountered in mythology. Then we would expect meditation, and perhaps prayer too, to be catalysts for synchronistic coincidences, the expressions of the activation of such archetypes.

Jung believed that synchronicity argues for a level of reality which, like Bohm's implicate order, forms a common foundation for both mental and physical existence. We will examine this foundation, for which Jung used the medieval term *unus mundus* (one world), in Chapter 4. Let us conclude this chapter by restating the possibility that an opening to this level of reality could be a significant step in the activation of psychological archetypes, and thus to synchronicity itself.

Hermes

In a room filled with conversation there occasionally spreads a lull of profound silence. Among the ancient Greeks it was said that at such moments Hermes had entered the room. Hermes symbolizes the penetration of boundaries—boundaries between villages, boundaries between people, boundaries between consciousness and the unconscious. With his winged sandals and cap of invisibility, Hermes brings the numinous power of the unconscious into the world of ordinary experience, carrying with it the capacity to catalyze synchronistic coincidences.

The room in question, however, is not located in a house made of wood and stone. It is the bicameral room of the human brain, located in the skull. And the conversation that fills it is the myriad voices of the mind. When these are silenced it becomes possible for archetypes to move forward toward expression. As we have seen, the activation of such archetypes is the cornerstone of synchronicity.

part two

SYNCHRONICITY AND MYTH

The first function of a mythology is to waken and maintain in the individual a sense of wonder and participation in the mystery of this finally inscrutable universe, whether understood in Michelangelo's way as an effect of the will of an anthropomorphic creator, or in the way of our modern physical scientists—and of many of the leading Oriental religious and philosophical systems—as the continuously created dynamic display of an absolutely transcendent, yet universally immanent, mysterium tremendum et fascinans, *which is the ground at once of the whole spectacle and of oneself.*

JOSEPH CAMPBELL
The Way of the Animal Powers, Volume I

A GOLDEN BEETLE:
carl jung and synchronicity

*To ascribe an intention to chance is a thought
which is either the height of absurdity or the
depth of profundity—according to the way in
which we understand it.*

ARTHUR SCHOPENHAUER,
*On the Apparent Design in the
Fate of the Individual*

JUNG AND SYNCHRONICITY

Paul Kammerer was a fascinating personality, and his first explorations of meaningful coincidences were a shining example of his unorthodox brilliance. Unfortunately, his approach did little to disclose the deeper meaning of the topic. A more intuitive approach was required, and such an approach was provided by Carl Jung.

Jung sensed from the beginning that meaningful coincidences were merely the surface effects of a deeper, more holistic reality. He saw in synchronicity a clue to a marriage between the essence of human nature and the external world of physical reality. He used the medieval term *unus mundus* to refer to this tacit unitary reality, the one world which must exist behind the poles of spirit and matter. He also found this marriage represented in the Chinese notion of Tao, an unmanifest unity that underlies, organizes, and patterns all that is manifest.

Living much of his life during the first half of this century, Jung was one of the first Europeans to develop an intense interest in Eastern thought, an interest nurtured by his friendship with the great sinologist Richard Wilhelm, the first entirely successful translator of the *I Ching*. Jung believed that this remarkable book of wisdom and divination, derived primarily from the fertile ground of Taoism, provided "one of the oldest known methods for grasping a situation as a whole and thus placing the details against a cosmic background."[100] He regarded many if not all methods of divination as efforts to tap, in a controlled way, the same process exhibited in everyday instances of synchronicity. (See Appendix I for a further discussion of divination.)

Jung felt that our connection with this holistic ground of reality occurs at the level of the unconscious mind. Thus, it is not surprising that he regarded meaningful coincidences as symbolic of unconscious activity. As indicated earlier, Jung himself introduced the term synchronicity, and by it he specifically meant coincidences which mirror deep psychological processes. His best-known instance of synchronicity, for example, is rich with meaning. This is the now-familiar case of the golden beetle, in which a local variety of scarab beetle was heard tapping at his window at the very moment a patient was recounting a dream about its sacred Egyptian relative. "I opened the window," Jung later wrote, "and caught the creature in the air as it flew in." He observed that "the scarab is a classic example of the birth symbol," suggesting the idea of self-transformation and rebirth. This is exactly what Jung's patient was struggling with at the time of her dream, and the appearance of an actual beetle was enough to break her psychological blocks and allow her to move toward her own transformation.[101]

Jung himself experienced many remarkable instances of synchronicity. During one twenty-four hour period in April 1949, he noted the recurrence of the fish theme on no less than six separate occasions. He described these in his book, *Synchronicity: An Acausal Connecting Principle*:

Today is Friday. We have fish for lunch. Somebody happens to mention the custom of making "April fish" for someone. That same morning I made a note of an inscription which read "Est homo totus medius *piscis* ab imo" [The inscription contained a figure that is half man and half fish]. In the afternoon a former patient of mine, whom I had not seen for months, showed me some extremely impressive pictures of fish which she had painted in the meantime. In the evening I was shown a piece of embroidery with fish-like sea monsters in it. On the following morning of April 2 another patient, whom I had not seen for many years, told me [of] a dream in which she stood on the shore of a lake and saw a large fish that swam straight towards her and landed at her feet. I was at the time engaged on a study of the fish symbol in history.[102]

Jung recorded these events while sitting by a lake. Finishing the last sentence, he walked over to the seawall where he saw a dead fish, about a foot long and apparently uninjured, which had not been there the previous evening. If we include this, we have a total of no fewer than eight fish-related incidents.

In the case of the fish, as with the golden beetle, the symbolic richness of the subject is apparent. It is, for example, a symbol traditionally associated with Christ, with the birth of a hero, and with the life and instincts of the unconscious. It sometimes appears as a fertility symbol. Jung's own involvement with the fish symbol is evident, as he himself tells us that he was engaged in the study of its meaning in history.

It is said that the idea of synchronicity came to Jung in the 1920s during a dinner conversation with Albert Einstein. It seems appropriate that such an idea would result from a conversation between two men who would contribute so greatly to a sense of unity in the cosmos, Jung in the inner realm and Einstein in the outer. As we have seen, however, the notion of meaningful correspondence between causally unrelated sequences of events goes back much further. Jung, himself, refers to Schopenhauer as the "godfather" of his ideas.

In 1850, Schopenhauer wrote an essay, "On the Apparent Design in the Fate of the Individual," dealing with the "simultaneity of the causally unconnected, which we call 'chance.' "[103] Schopenhauer believed that each individual life follows a preestablished pattern or fate. He likened the chain of causality that establishes the path of a single person's life to a meridian on the globe, while the lines of latitude represent interactions between lives of separate individuals:

> The fate of one individual invariably fits the fate of the other, and each is the hero of his own drama while simultaneously figuring in a drama foreign to him—this is something that surpasses our powers of comprehension, and can only be conceived as possible by virtue of the most wonderful pre-established harmony. [104]

Schopenhauer himself no doubt owed a debt to Leibnitz, who wrote that the soul was "a perpetual living mirror of the universe." This latter notion embraces the alchemical idea that the human soul contains a mirror reflection of the entire cosmos in miniature, an idea that was in turn related to the medieval notion of sympathies, described in the Introduction. All such concepts point by implication to a state of unity, a unity familiar to mystical thought throughout the world. In Mahmun Shabistari's words:

> The universe is contained in a mosquito's wing. . . .
> But if one atom from its place is moved,
> The universe at once is overturned. [105]

Jung was impressed with Einstein's calculation of the staggering amount of energy that lies hidden in the atom. Indeed, during the first half of this century the whole world was amazed by this, and particularly by the quantities of energy that could be released by "splitting" the atom. Jung wondered if there might be a metaphor here for the human psyche, a structure which also enclosed complex and hidden dynamics. Might not it likewise contain vast, concealed energies which in the right circumstances could be released? Here we touch the core of the idea of how synchronicity operates psychologically. To go further, we must know more about the nature of the psyche itself as Jung conceived it.

THE PSYCHOLOGICAL ROOTS
OF SYNCHRONICITY

The fact that synchronistic coincidences may express a common form and meaning in both human consciousness and in the physical world, as was the case with the scarab beetle, implies that the origin of such coincidences stands behind both of these realms. This suggests that synchronicity is rooted in the deepest level of the mind or psyche, the *unus mundus*. Jung referred to this level as a pseudopsychic or *psychoid* state, since it is not strictly psychic but partly physical as well. The psychoid state lacks articulation, representing a fusion of both inner and outer reality. In it coexist all past and future possibilities. Out of it come those exotic processes that Jung is famous for, the archetypes. These are concentrations of psychic energy, universal in their essence, which manifest as particular themes or motifs that emerge from the unconscious—Jung used the term "collective unconscious" to refer to the universal (collective) aspect of this level of the psyche to exert a dramatic influence on conscious experience and behavior. Jung believed that the activation of an archetype is what triggers a synchronistic coincidence.

An archetype is the potential for a particular theme or image which lies dormant until triggered by some situation in the environment or in the conscious or unconscious mental life of the individual. Then it resonates like a bell, its chords heard and felt throughout the personality. Archetypes can be experienced in oneself or projected onto others. An example is the Hero, embodied in figures such as the Arthurian knight Sir Lancelot or the hero of a cowboy movie or a detective story. When such images touch and excite us it is because they resonate with this archetype. Archetypes can be projected onto real persons as well, as a photograph is projected onto a screen. In this case we see those persons as bigger than life. We can also personally identify with an archetype, becoming, in a sense, possessed by it. Jung referred to this as "inflation." A person strongly possessed by an archetype seems blown up bigger than life, as if puffed up with hot air! Such persons may actually dream of rising into the atmosphere, unable to keep their feet on the ground.

Another archetype is that of the Wise Old Man, the embodiment of deep and ancient wisdom personified in literary and film characters such as Merlin the magician, Gandalf the Gray, and Obi-Wan Kenobi of *Star Wars*. Each wields magic powers that derive from his mastery of ancient, all-but-lost knowledge. Other examples of less mysterious and more beneficent wise old men, such as the wise men from the East, touch upon another archetype, that of the God-Man, or *manna man* to use Jung's term. This is the ideal of a human embodiment of the essence of the divine. Projecting this image onto someone else is to give that person great emotional power over yourself. Needless to say, this can be very dangerous unless that person is a remarkably worthy individual. To identify personally with this archetype is a major obstacle to inner growth, for it virtually guarantees an absence of humility. It is fine for others to refer to Mohandas Gandhi as *Mahatma*, "the great soul," but beware of those who confer such titles upon themselves.

An archetype may take the shape of a person such as the hero Sir Lancelot, but at root archetypes are amorphous and can assume an endless variety of forms. In Chapter 2 on physics, for instance, we saw that the Great Mother archetype is represented not only in the myriad figures of Earth goddesses but also in images from nature such as the bear, the rabbit, the cow, a plowed field, and a flower.

Jung believed that at the center of the psyche, the *unus mundus* underlies both the inner world of the psyche and outer world of objective reality. Here, in the tradition of Schopenhauer and especially of Leibnitz, Jung viewed the inner world of the psyche as a mirror of the outer world, referring to the deep unconscious as a "collective living mirror of the universe."[106] At the psychoid level, the outer and the inner worlds are fused in fundamental unity. Thus, archetypal expressions arising from this center carry not only inner but also universal truth. When we meet such expressions in our own lives, we experience facets of reality that are mythic in scope and significance. Let us examine further the notion of myths.

Myths and Archetypes

We believe, as Jung did, that archetypes are the foundation of myths, which are essentially plots projected out from the unconscious. People commonly use the word "myth" in a variety of ways. On the most superficial level, a myth simply means a statement that is not true. For instance, it is a myth that men and women enjoy equal employment opportunities. On another level, myth refers to an imaginative, often romantic narrative which may express some emotional truth but is not based on fact. This is often the way we understand the Greek myths.

As mythologist and historian William Irwin Thompson points out, myths on a deeper level are narratives that provide answers to three essential questions: What are we? Where do we come from? Where are we going?[107] On this level, myth is synonymous with exoteric religion as we usually think of it. The story of the Christ, for instance, lends meaning to the life of the devout Christian. Here, the question of the historical, "factual" nature of the myth is not central to its meaning and function. Greek myths were no doubt experienced on this level by many of the early Greeks themselves. Much of the serious study of mythology occurs at this level. In the *Odyssey*, for example, one might interpret Telemachos' search for his father Odysseus as the prototypical search of a young man for his profession, or on a larger scale the *Odyssey* itself as a journey into the male psyche. Or one might interpret the virgin birth of Christ as witness to his purity. These are essentially symbolic interpretations that allow particular stories with all their detail and color to speak to us in our own lives.

Interestingly, the unique historical events of the life of an individual, while profoundly important to the development of that life, may at the same time take their form from the template of a much broader truth. Consider Christ, whose individual life takes its form as an expression of the archetype of the Divine Son. Joseph Campbell recounts another example of a particular life representing a universal core of meaning in the legend of a Siberian shaman named Aadja. After his death as a young man, Aadja was reborn into the spirit world where, following various adventures, he was placed at the topmost level of the great world tree, there to be suckled by a white-winged reindeer. From this vantage point, he witnessed another shaman's visits to the spirit world to retrieve lost or stolen souls. Eventually Aadja was reborn of a human mother. Campbell notes that two points of view must be kept separate in understanding such legends as this, one historical and one symbolic:

> In the religious lives of the "tough-minded," too busy, or simply un-talented majority of mankind, the historical factor preponderates. The whole reach of their experience is in the local, public domain and can be historically studied. In the spiritual crises and realizations of the "tender-minded" personalities with mystical proclivities, however, it is the non-historical factor that preponderates, and for them the im-agery of the local tradition—no matter how highly developed it may be—is merely a vehicle, more or less adequate, to render an experience sprung from beyond its reach, as an immediate impact. For, in the final analysis, the religious experience is psychological and in the deepest sense spontaneous; it moves within, and is helped, or hindered, by historical circumstance.[108]

Campbell is saying that mythological experience confronts indi-viduals through the vehicle of historical or local forms or images, yet these experiences have a reality beyond the accident of a particular form. The death of Aadja, his ascent into the spirit world and return to the world of humanity may have been an actual visionary expe-rience, but its meaning—that one may be transformed by death, nurtured by the power of the spirit, and reborn to humanity—could have been expressed in many ways. Such mythological experiences are truly archetypal: they express a psychological reality which cannot be denied by reference to a historical level of understanding.

Ironically Western science enters the realm of myth when it tries to answer Thompson's three fundamental questions. In this sense, Freud and Darwin were mythmakers as much as Dante and Milton. In a larger sense, the whole attitude of mechanistic science since the time of Galileo, Descartes, and Newton is mythic, telling us that the cosmos is a vast machine, of which the individual is but an insignificant part.

In tracing what we mean by myth, however, we find that there is a still deeper level which is the wellspring of all true mythology. It might be termed the dramaturgical level, because in it myths are actually played out in the lives of individuals, as if on a stage. The example of Christ's life or Campbell's story of Aadja, while they may be understood as symbolic only, also may be understood from this point of view. Here, mythic themes become living realities. At the previous level one empathized with the myth and perhaps even interpreted it, but at this level one lives it. For the individual caught up by the power of an archetype, the symbolic meaning of his or her life is no abstraction but a powerfully felt and utterly convincing reality which, with or without the individual's conscious participation, directs and forms the nature of the world.

In his clinical practice, Freud came across numerous instances of early childhood memories that suggested an intense attraction of children roughly two years of age for the parent of the opposite sex. He termed this the Oedipal situation because it reenacted the essentials of the old Greek myth dramatized in Sophocles' play, *Oedipus Rex*, in which Oedipus unknowingly marries his own mother after slaying his father. The correspondence between the mythic tale of Oedipus and the situation in early childhood was, in Freud's mind, no accident. The story unfolds a drama proper to the inner psyche. In Jung's terms we would say that the inner drama is archetypal, projected into the outer world as a myth. When we actually live the myths, we experience them at the dramaturgical level.

When we examine our lives for mythological content, we often can see that various aspects of our personalities seem to correspond to different mythological figures. This is an essential notion in Jungian thinking, where mythological personages, most often gods and goddesses, are frequently identified as expressions of archetypes. We will have more to say about this in Chapter 6, but note here that people can easily come under the influence of such mythic or archetypal figures, causing them to behave as if they were living a part in a mythological drama. For instance, in a discussion of defensive styles, storyteller Michael Meade observed:

You could have from the feminine point of view, as a style of self protection, the boundary keeping of Artemis. Artemis is a virgin goddess and she stays a virgin for the most part. What virginity means is a style of self protection, keeping intact. Artemis' style of keeping intact is through communication with nature. For instance, Artemis has the manner of protection of the deer, the way the deer home their territory. Quick movements. Alertness to things that might be penetrating one's boundaries from any side. Artemis carries a bow and arrow, so when you go too far into the territory of Artemis you get hit with an arrow. Some women can do that—I'm taking a woman as an example because Artemis is a feminine figure—you say something to a woman and she turns and hits you with a darting stare. That's Artemis sending an arrow out and you get the message right away, "back off."[109]

Many such examples could be given for both men and women, and indeed it seems to be the case that we can discover much about our ourselves, the mythic characters we play, and the plots we live out daily, by examining our lives in terms of mythological dramas.

Acausal Connections

In Jung's thinking, the activation or awakening of an archetype releases a great deal of power, analogous to splitting the atom. This power, in the immediate vicinity of the psychoid process from which the archetype takes its origin, is the catalyst for the synchronistic event. There is a buildup of energy at this most primitive and undifferentiated level of the personality. The energy overflows, triggering the synchronistic event. The idea is that the activation of an archetype releases patterning forces that can restructure events both in the psyche and in the external world. This restructuring proceeds in an acausal fashion, operating outside the laws of causality.

The power that is released is felt as *numinosity*—literally a sense of the divine or cosmic. It is described by Jung's student Ira Progoff as "a sense of transcendent validity, authenticity, and essential divinity."[110] Numinosity is described in the *Iliad* when the old Trojan king, Priam, secretly crosses the Greek battle lines to plead with Achilles for the return of the body of his dead son Hector. Priam is approached by the god Hermes disguised as a young man:

The old man was dumbfounded and filled with terror; the hairs stood up on his supple limbs; he was rooted to the spot and could not say a word. But the Bringer of Luck [Hermes] did not wait to be accosted. He went straight up to Priam, took him by the hand.[111]

As we have seen, Hermes, who is referred to here as the "Bringer of Luck" was, of all gods, the one most identified with synchronicity. We will have much more to say of him and the archetype he represents in Chapter 5.

Jung developed his ideas of the acausal nature of synchronicity in close collaboration with the quantum physicist Wolfgang Pauli. Pauli is famous for predicting the existence of a strange and important subatomic particle, the neutrino, in 1930, twenty-six years before its existence was actually confirmed in the laboratory. A more fundamental contribution of Pauli's is the exclusion principle, which states that no two electrons can occupy any planetary orbit within the atom. This principle, though no more than a formal mathematical characteristic of the equations regarding the principle of symmetry, allows quantum physics to make connection with most of the pragmatic properties of the real world. These include the physical bases of chemical bonds and the solidity of matter.

Arthur Koestler observed that it is not surprising that the person who discovered the exclusion principle, which "acts like a force although it is not a force," would be one of the first to realize the limitations of conventional science. Indeed, Pauli and Jung proposed that the traditional triad of classical physics, space, time, and causality, be supplemented by a fourth element, synchronicity, producing a tetrad. This fourth element operates in an acausal fashion, representing the polar opposite of causality. Pauli and Jung drew this polarity at right angles to space and time (see Figure 1 on page 76). Pauli actually believed this acausal process to be "metaphysical," operating at a more fundamental level than the laws of physical causality.

As we have seen, Pauli was no stranger to synchronistic coincidences. He was well known for the Pauli effect, by which delicate scientific instrumentation seemed inevitably to break down in his

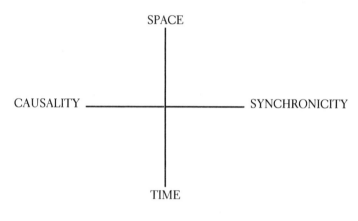

Figure 1

presence. The term is a humorous takeoff on the Pauli exclusion principle. Pauli's personal life as well was characterized by rich networks of coincidences, particularly during periods of crises.

JUNG AND KAMMERER

Coincidences rich in symbolic meaning, so characteristic of those reported by Jung, seem at first glance to contrast sharply with the those reported by Kammerer, which typically dealt with rather mundane facts such as the recurrences of numbers, letters, or names. Taken together, they could fill a curiosity shop with interesting but apparently meaningless tricks of chance. Jung himself was critical of such coincidences and in most instances did not consider them to be true synchronicity.

Jung was explicit in limiting his notion of synchronicity to those coincidences that express symbolic or mythic meaning. In a psychological sense, each synchronistic event was thought to involve two mental states at once. One was the ordinary state of mind, the product of whatever activities the person was involved in at the moment. The other was a nonordinary state that resulted from the activation of an archetype. The latter was largely unconsciousness but carried with it the feeling of numinosity or absolute cosmic authority. This feeling was not always overwhelming but was always present.

Kammerer, for his part, must certainly have felt more than a little excitement with the coincidences he reports in his book, *Das Gesetz der Serie*. After all, he kept a log of them for many years, starting at the age of twenty. Perhaps we should not be too quick to dismiss these coincidences. The very presence of Kammerer's intense fascination strongly suggests the proximity of unconscious activity— activity that may have been more than casually related to the coincidences themselves.

Interestingly, Jung considered just such an intense interest critical to the function of ESP, which he believed to be due in large part to synchronicity. Jung was intrigued by J. B. Rhine's well-known studies of the paranormal, conducted at Duke University. Rhine was often successful in demonstrating phenomena such as telepathy, precognition, and telekinesis but was regularly beset with problems in maintaining high levels of performance from his subjects. No matter how well they performed in their initial trials, their performance would gradually deteriorate to the chance level over a period of days, weeks, or months. Rhine was never able to account for this, but Jung's idea was that it had to do with a loss of interest. Most ESP experiments are mundane, for example, guessing the outcomes of long successions, perhaps hundreds, of coin tosses before they are actually thrown. Under these circumstances it is difficult not to become bored and lose interest as time goes by. Such a loss of interest reveals a loss of involvement of the psyche as a whole, and thus an increasing failure to engage the deeper levels of the unconscious.

Kammerer evidently experienced no loss of interest or involvement. His intensity seemed to serve as a kind of psychological catalyst for the synchronistic coincidences that he observed. Along the same lines, Alan Vaughan, in his book, *Incredible Coincidence*, notes that when one becomes interested in synchronicity, meaningful coincidences begin to appear on every street corner! He humorously terms this phenomenon the "synchronicity of synchronicity," saying that, for him, lecturing on synchronicity almost always provokes a few synchronistic episodes. He notes that the synchronicity of synchronicity seems to affect virtually everyone who becomes interested in the topic.[112]

Kammerer was enthusiastic, sincere, and original in his work, but his analytic and statistical approach relegated his findings to the status of the merely exotic. Jung's intuitive and holistic approach, on the other hand, stands the test of time in its rich and thought-provoking placement of synchronicity at the core of the meaning of our lives.

· 5 ·

HERMES THE TRICKSTER

---

*With considerable justice, it can be said that
the Greeks named the experience of synchronicity "Hermes."*

MURRY STEIN
In Midlife

The themes carried by archetypes
are universal: they are neither wholly internal nor wholly external
but are woven into the deepest fabric of the cosmos. This notion
is supported by Jung's idea that archetypes have their origins in
the *unus mundus,* or "one world," which is at the foundation of the
psyche and the objective, physical world. Bohm's concept of the
holographic universe offers similar possibilities. It follows, then, that
myths as expressions of archetypes might be expected to portray certain
aspects of the objective world as well as depicting psychological realities. Indeed, many of the Greek gods represent aspects of reality that
overarch both the inner worlds of human experience and the external
worlds of nature and society.

Zeus, for example, literally means "light" or "shower of light." In
the world of nature, Zeus was associated with the bright sky and was
seen as the source of atmospheric phenomena. Perhaps the most
dramatic of these are lightning and thunder, which are unusually

common in certain of the mountainous areas of Greece and with which Zeus was particularly affiliated. As the central figure in the religion of Homeric Greece, however, Zeus symbolized the inner experience of light and illumination. His illumination kindles the spirit of lucidity that permeates the other gods, and indeed the entire ancient Greek culture. It is mirrored in their art, poetry, and philosophy. The great classicist, Walter F. Otto, has shown that the other Greek gods can be seen as outward projections of the numerous aspects of Zeus.[113]

Understanding Zeus, we come to understand something of the Greek experience. We also see that as myth, Zeus represents valid aspects of both inner human consciousness and outer atmospheric phenomena. Both are as real today as they were to the Homeric Greeks. Thus, the myth is a signpost pointing to actual features of the real world of mental and physical events.

We have already noted that many fundamental notions in science have mythological predecessors. The idea of causality, for instance, can be traced to mythology. It reflects the ancient belief in a rational and ordered universe. In modern times this notion was expressed by Einstein in his famous comment, "God does not play dice with the universe." Marie-Louise von Franz points out that the notion of the atom is based, to use Jung's words, "on the 'mythological' conception of smallest particles," the soul-atom.[114] Atomic theories were developed in the classical world by Leucippus and Democritus, but the essential concept of the atom has found expression through sources as widely separated as the Aborigines of central Australia and the ancient middle eastern Gnostic belief in light-nuclei that contain the potentiality of all that is or can be. Dieter Mahnke, in a study of the imagery of the sphere in Western traditions, has extensively documented the idea that prior to modern physics virtually all conceptions of the atom, as well as ideas about space and time, "in the end, derived from a mandala-formed God-image."[115] The latter, as Jung has shown, is also symbolic of the archetype of the self, representing the unity of the personality.

Of special interest is another aspect of nature that is seen both in the realms of myth and of objective reality, in the inner world of the psyche and the outer world of external events. This is synchronicity itself, the uncanny intrusion of the unexpected into the flow of commonplace happenstance, an intrusion that hints at an undisclosed realm of meaning, a disparate landscape of reality that momentarily intersects with our own. The seeming intelligence with which synchronistic events orchestrate themselves gives the sense that some personal but larger-than-life agency is operating behind them. We might imagine, for instance, that some whimsical god has taken a personal interest in our affairs and arranged them in some pattern understandable, perhaps, only to himself. This god was known to the ancient Greeks as Hermes; he is the topic of this chapter. But let us begin with the unexpected.

No realm of human experience is free from intrusions of the unexpected, sometimes carrying with them the promise or the threat of revolutionary change. Physicist and historian of science Thomas Kuhn, in his famous examination of scientific revolutions, points out, for instance, that the history of science is characterized by long periods of uneventful investigation of predictable and well-understood phenomena, occasionally punctuated by the appearance of utterly unexpected and unaccountable observations.[116] These "anomalies" can in time lead to the overthrow of entire paradigms of scientific knowledge. Such an anomaly was the first observation of X rays by Wilhelm Röntgen near the turn of the century, as was the data which disproved the so-called ultraviolet catastrophe in thermodynamics. The resolution of these led, in large part, to the discovery of the entire world of quantum physics.

Synchronistic coincidences, such as the entry of the beetle into Jung's study at just the right moment, likewise share the quality of being unexpected. They stand out from the background of everyday events because of the sense of purpose or meaning that accompanies them. At the same time, they violate our confidence in a world of events chronologically ordered and based on cause and effect. They create a conspicuous discontinuity in ordinary reality, an opening to the miraculous.

In the mythologies of many peoples, the mythic figure who is the embodiment of the unexpected is the Trickster, who steps godlike through cracks and flaws in the ordered world of ordinary reality, bringing good luck and bad, profit and loss. The trickster god is universal. He is known to the Native American peoples as Ictinike, Coyote, Rabbit and others; he is Maui to the Polynesian Islanders; Loki to the old Germanic tribes of Europe; and Krishna in the sacred mythology of India. Best known to most of us in the West is the Greek god Hermes, who represents the most comprehensive and sophisticated manifestation of the Trickster. Homer calls Hermes the "Bringer of Luck." He is also known, in one of the many paradoxes that characterize Hermes and other trickster gods, as the patron of both travelers and thieves. He is the Guide of Souls to the underworld and messenger to the gods. As all these roles suggest, he is the quintessential master of boundaries and transitions. It is by this mastery that he surprises mundane reality with the unexpected and the miraculous.

THE BOUNDARY DWELLER

As the master of the unexpected, Hermes performs his magic by virtue of his command of boundaries and his ability to cross them effortlessly. Where a situation is defined by boundaries or boundary crossing, Hermes' presence is natural. He is the god of thresholds and transitions. In Chapter 4, for instance, we saw that in the *Iliad* Hermes appears to old King Priam as the king approaches the boundary of the Greek encampment in an effort to reclaim the body of his dead son Hector. Hermes comes to Priam as a young man, a guide, who leads him safely into the enemy camp, then out again. In the *Odyssey*, Hermes appears before Odysseus when, on Circe's island, he approaches the witch's dwelling in search of lost crew members that she had turned into swine. The god in the form of a youth shows

Odysseus a magic plant which protects the Hero from Circe's charms.

In both of these instances the Trickster bestows good fortune. Indeed, Hermes has been called "the friendliest of the gods to man".[117] These two episodes, however, reveal a more basic aspect of the god, his appearance at boundary or threshold situations; indeed, Hermes is the god of thresholds, not only physical thresholds but, more importantly, thresholds between states of human experience: between day and night, sleeping and waking, consciousness and the unconscious, life and death.

Karl Kerenyi, the classical scholar who has done the most to enrich our understanding of Hermes, explains his effect at psychological boundaries as *Hermetic journeying*. Whereas ordinary traveling simply involves physical movement from one place to another, the Hermetic voyager enters a boundary zone, a liminal space between ordinary states of experience. "In reality, he [the traveler] makes himself vanish ('volatizes himself') to everyone, also to himself."[118] This is the deeper meaning behind the presence of a herm, the stone block representations of Hermes found at entrances to homes and along ancient highways. His presence at entryways and gates in particular suggests the condition of liminality, as do various inscriptions written on doorways and dedicated to him. Kerenyi comments:

> In his official capacity as mediator between the worlds of night and day, spirits and men, and (standing before the temple) between the worlds of Gods and mankind, he is called *Propylaios* ("before the gate"). . . . One inscription names him *Pylaios* ("the one at the entrance") and *Harmateus* ("the driver of the chariot"). Two other epithets— *strophios* ("standing at the door-post," also "cunning," "versatile") and *stropheus* (the "socket" in which the pivot of the door moves)—show him closely related to door hinges and therefore to the entrance, but also to a middle point, to the socket, about which revolves the most decisive issue, namely the alternation life-death-life.[119]

Because he is at the "pivot point" between the transformations from life to death and back again, Hermes is at just that psychological nexus where transformation occurs, where for example, as we are changed by new facts or experiences, we undergo a transformation into a different person—a symbolic death and rebirth. Wherever human experience breaks through frontiers to the unexpected or undergoes transitions, there we may find the archetype represented by Hermes or by another of the trickster gods.

Synchronistic coincidences are, from the Jungian perspective, boundary events. They manifest, for instance, as transitions across the margin between psychological reality on the one hand and physical reality on the other. The arrival of the beetle at Jung's window, as well as the various representations of fish Jung himself experienced while working on the meaning of the fish symbol, can be seen as translations into the material world of psychological actualities. Such coincidences, like dreams, also carry symbolic messages across the boundary of the unconscious into consciousness. As with dreams, their meaning may not be apparent, though they still may transmit a strong emotional message. Upon seeing the actual beetle fly into the consultation room, for example, Jung's patient was so emotionally jarred that she was able to begin to free herself of a neurotically rigid worldview.

We seem most accessible to the synchronistic gifts of the Trickster when we ourselves are at or near boundaries or are experiencing transition states. For example, meditation, which carries consciousness beyond its ordinary boundaries, seems to catalyze meaningful coincidences. One effect of meditation is to soften the barriers between the conscious and unconscious.[120] Transpersonal psychologist Ken Wilber suggests that, in addition, meditation gradually elevates consciousness to a point at which "various high-archetypal illuminations and intuitions occur."[121] If this is the case, it is not surprising that meditation tends to promote the activity of archetypes, which in turn lie at the root of synchronicity.

In the experience of the authors, traveling, especially by public transportation (a plane, bus, or train) is also a strong catalyst for synchronicity; not only of chance encounters with other persons, which are often remarkably meaningful, but also of "accidental" discoveries of books, magazine articles, and so on. Traveling is a transition in physical space, one that is also accompanied by a transition in one's state of mind. We leave one environment, perhaps home, to travel to another—perhaps a business meeting or a vacation—that carries a distinctly different mood. While we are physically crossing geographic boundaries, we are subjectively making a transition from one mental atmosphere to another. The whole experience of being on a trip, particularly if one is not a seasoned traveler, carries a psychological sense of transition, expectation, and openness to new experience.

Similarly, periods of major life transitions seem to be occasioned by an abundance of meaningful coincidence. Personal growth seems not only to facilitate synchronicity, but in turn to be facilitated by it. In his book, *In Midlife*, Murry Stein notes that the period of the midlife transition, or *midlife crisis*, is visited by more than its share of synchronicity, and further that the patron of the this transition is Hermes himself.[122] The story of Hermes' appearance as guide to King Priam, who seeks the body of his son Hector, is symbolic of the midlife search for the corpse of lost heroic youth. It is necessary to bury this corpse—to put it to rest—in order to get on with the business of the latter half of life. This latter period, in Jung's thinking, is a time for fulfilling one's unique calling in life.

Other major life transitions, such as career changes, may be similarly visited by frequent and dramatic synchronistic episodes. In *The Luck Factor*, Max Gunther reports that his career as a writer began at *Time* magazine with a job that he got because he happened to be on the right street corner at the right time, where he met the right person who offered him the job.

The most dramatic transition of all is death. Here we find Hermes in the role of *psychopompos*, literally the "spirit who shows the way," or Guide of Souls to the underworld. Over the years many observers have noted that no other event in human experience is associated with so rich an array of psychic phenomena as is death. Synchronicity is no exception. There are few who could not tell some story from their own family histories that relates a meaningful coincidence to the death of a loved one. Many stories have to do with omens, synchronistic coincidences that seem to foretell the death before it happens. The second example that Jung recounts in his original 1952 essay on synchronicity, in fact, concerns such an omen. It is about a woman who had witnessed large numbers of birds gathering outside her house at the deaths of her mother and grandmother. Her own husband's death involved a similar incident, and came in a most unexpected way. He was one of Jung's own patients and was completing his session when Jung noticed some apparently innocuous symptoms which seemed to him, with his medical training, to represent possible heart disease. Jung referred the man to a specialist who found nothing wrong with him. In the meantime, his wife at home was becoming increasingly alarmed by a flock of birds that had gathered on the house. On his way home the man collapsed, his medical report in his pocket, to be taken home to die.[123]

In considering the symbolic significance of the birds, Jung observes that in ancient Babylon the souls in the underworld wore a "feather dress," and that in Egypt the *ba*, or soul, was thought of as a bird. In Homer and other classical sources the souls of the dead are said to "twitter" and to flutter about. Such examples and many others suggest that, in certain contexts, birds symbolize death or the departure of the soul. Interestingly, omens frequently involve natural signs and events as opposed to the perhaps more frequent synchronistic coincidences which consist of human creations such as numbers, words, books, or ideas. (Consult Appendix I for more about omens.)

HERMES AND THE IMAGINATION

A colorful Trickster appears in the stories of the American Plains Indians, where he is well known as the unpredictable Coyote. Here he plays a central role in the mythic order of the world. In some stories he actually is said to have created it. But the Trickster is nothing if not paradoxical, and so he is also a joker, as selfish and unreliable as they come. His faults are often ridiculously evident:

> Coyote taught the people how to eat, how to wear clothes, make houses, hunt, fish, etc. Coyote did a great deal of good, but he did not finish everything properly. Sometimes he made mistakes, and although he was wise and powerful, he did many foolish things. He was too fond of playing tricks for his own amusement. He was also selfish, boastful, and vain.[124]

Coyote is at once a clown and a creator, gift giver and thief. Above all, he mocks and disrupts convention, order, and preconception.

In his pranks, his thievery, and his disdain for convention, Coyote is very much like Hermes, especially in the latter's childhood. When Hermes was born:

> His mother Maia laid him in swaddling bands on a winnowing fan, but he grew with astonishing quickness into a little boy, and as soon as her back was turned, slipped off and went looking for adventure. Arrived at Pieria, where Apollo was tending a fine herd of cows, he decided to steal them. But, fearing to be betrayed by their tracks, he quickly made a number of shoes from the bark of a fallen oak and tied them with plaited grass to the feet of the cows, which he then drove off [walking backwards] by night along the road.[125]

Apollo eventually trailed the little thief to the cave where he had wrapped himself again in his swaddling clothes. There he accused him of the theft of his cattle. Hermes defended himself, claiming that he was too young to have stolen the cattle (he was born that same day!). Unconvinced, Apollo began to reproach him. Picking up the mischievous infant, however, he received a reply in the form of what *The Homeric Hymn to Hermes* calls "an Omen, an evil belly-tenant"; in other words, Hermes broke wind, a decidedly un-Olympian reply to the aloof and majestic Apollo.

In this story, not only does the infant Hermes violate the taboo against interfering with Apollo's cattle but absolutely violates the seriousness and respect due to great Apollo by passing gas practically in his face (in disgust, Apollo drops the infant).[126] Yet there is more to the story. After stealing the cattle, Hermes had sacrificed two of them to the twelve Olympian gods—which now included himself. In order to do so he invented both sacrifice and fire, and he showed true Olympian restraint and consciousness in taking only the portion of the sacrifice that rightly belonged to him, in spite of his extreme "greediness for meat." This episode beautifully illustrates how the Trickster, in performing his pranks, becomes also a creator of culture, here in the form of sacrifice and fire. We will soon learn more about the Trickster's role in the bringing of fire.

Hermes, Coyote, and the other trickster gods are filled with irreverent vitality and creativity. In this they embody the life-giving power of the human imagination. Otto observes that Hermes "mysteriously bobs up everywhere."[127] Like other tricksters, the imagination knows no boundaries and may appear anywhere. As patron of travelers and guide of souls, Hermes, who personifies the imagination, leads us to the heights or depths of experience, to the light of Olympus or to the shadows of Hades. He also guides us across the boundaries of ordinary reality to experience other states of consciousness.

For instance, it is Hermes who symbolically conducts our nightly transition to the dream world. In doing so he reenacts his timeless role as guide of souls to Hades. In the underworld of dreams, events are seen from a perspective that is reversed from that of the daytime world, as if we were looking at our lives from behind the stage. Jungian analyst James Hillman, in *The Dream and the Underworld*, notes that the Egyptian underworld was literally upside down to the ordinary world, with its inhabitants walking on their heads.[128] From this altered perspective we can obtain insights into the nature of our personal lives and the problems that we have in day-to-day living. Carl Jung expressed this reversed aspect of the dream world in terms of compensation: dreams compensate for inadequacies in the way we see the world when we are awake. They show us what is missing.

Suppose, for instance, that during the day I argue with my wife, trying to "get her goat." At night, I dream that she and I are a two-person team in a foot race, and I push her off the track. It dawns powerfully upon me in the dream that I have just defeated my own team. Perhaps the next day I will feel more cooperative toward her, even if I don't retain any more than the residual feeling caused by the dream. That feeling, of defeating my own team by spiting my wife, is the inverse of what I felt the day before during the argument. It is what was missing, and it is what was needed to mend my relationship with her. Of course, the more I consciously work with the dream in order to understand its meaning, the more likely I will be able to correct what is inadequate in my attitude toward my wife.

Hermes was considered by the Greeks to be both an Olympian, filled with life and far removed from the world of the dead, and Hermes Chthonios, the epithet under which he was worshipped during the ancient Greek Anthesteria or all-souls festival. The title Chthonios—meaning "from under the earth"—indicates that Hermes also belongs to the underworld. Like the imagination, he can manifest in Hades as well as on Olympus, or at any point between. Flying on winged sandals, he can take us to the height of inspiration or the depth of depression.

Identifying the trickster Hermes with the imagination means that we recognize him as a world maker.[129] Indeed, it is the imagination that creates the worlds we live in, no less than Coyote is said to have created the world of the American Plains Indians. Further, the Trickster is associated with the light of consciousness in the form of the archetypal bringer of fire. The infant Hermes invents fire, and Joseph Campbell points out that myths of fire stealing from the world over show the thief of fire to be a trickster in one or another of many manifestations, whether it be the North American Coyote, the ancient Greek Prometheus, who shares with Hermes certain trickster qualities, or Kingfisher of the Andamanese, a primitive tribe who live on a remote island in the Bay of Bengal.[130] The association of Hermes with consciousness is also represented by the image of the god's head, which topped block-shaped herms and by his restraint in the face of a titanic appetite for meat in his sacrifice of Apollo's cattle. This light of consciousness is central to human creativity, for without it the productions of the imagination, if indeed they were possible at all, would find no expression. Indeed, as world maker and as fire bringer Hermes makes human life possible.

In his discussion of Hermes' chthonic origins, Kerenyi goes so far as to say that we come from the same place he does: "the same dark depth of being." Given this identity of origins, Hermes "creates his reality out of us, or more properly through us, just as one fetches water not so much out of a well as through the well from the much deeper regions of the earth."[131] What better way of saying that as an archetype, the Trickster, the boundary dweller, finds expression through human imagination and experience?

Hermes' central creative role is confirmed again in the image of his caduceus, or herald's staff, with its intertwined and copulating snakes.[132] These snakes are an image imported from the Middle East that represent, according to Campbell, "the monster serpent and the great goddess as serpent both renovating the world."[133]

Hermes' connection to the Great Mother goddess is developed in Kerenyi's discussion of Hermes' ancient association with the Ka-bierean mystery religion of northern Greece. Hermes probably came into being as an expression of masculine creativity, out of Brimo, the Great Mother goddess of that region.[134] The ithyphallic repre-sentations of Hermes—the phallus usually found on the front face of the stone block herms—indicates his essentially masculine and creative nature. Hermes, as an expression originating from the great goddess, is connected in the most intimate way to the ultimate source of all life, long represented by that goddess. Hermes, however, is no mere adjunct to the Great Mother but a separate creative force, different from the feminine, which finds unique expression in the particular forms that Hermes or the other trickster gods take. He is not a fertility god who serves the goddess but a force for creation in a masculine way—through the phallus, which symbolizes assertion and the penetration into the world in order to make life happen. The trickster god asserts his presence and power by entering into life in his own distinct way, and in so doing he makes that life happen; he makes it possible.

Hermes' creativity involving his theft of Apollo's cattle, initially a selfish act, ultimately leads to the invention of the lyre (from a tortoise-shell Hermes finds on the way); the invention of fire; and the beginning of sacrificial offerings to the Olympian gods and goddesses. The Trickster is credited with the introduction of marital confidentiality (the sanctity of the marriage bed) in one African tribal culture and the invention of storytelling in many cultures the world over.[135]

If, through the agency of imagination, the Trickster is the creator of life as we know it, of the world growing, regressing, changing all around and within us all the time, then Hermes and the other Tricksters are also connected to storytelling, and so to mythmaking. It is myth that gives us the meaning and indeed the order and structure of the reality we experience. Myths quite simply are stories or plots, to use Aristotle's term, by which we know the world and ourselves. Joseph Campbell comments, "Like dreams, myths are productions of the human imagination. Their images, consequently . . . are, like dreams, revelations of the deepest hopes, desires and fears, potentialities and conflicts, of the human will."[136]

Such notions are in surprising agreement with several current ideas in the brain sciences. In particular, Gordon Globus has recently proposed that the human brain operates in a holograph-like manner, generating something resembling its own internal holomovement or implicate order.[137] The basic notion is that the immensity and richness of the brain itself holds the potential to create all possible experiential realities. Dreams and waking life, for instance, are experienced as separate types of reality, as are various mystical and altered states of consciousness. The final choice about which is experienced at a given moment results from the imagination interacting with the limits placed on it by our mental state, the physical body, and the objective world.

The extent of the virtually unlimited reach of the imagination is suggested by the trickster god's paradoxical attributes or roles, such as being at once the patron of both travelers and thieves. The Trickster seems not to recognize any limited aspect of reality. Like the imagination itself, he moves in a "divine sphere of operation" which "is no longer delineated by human wishes but rather by the totality of existence. Hence it comes about that [his] compass contains good and evil, the desirable and the disappointing, the lofty and the base."[138]

One finds these paradoxical qualities especially highly developed in Hermes' medieval counterpart, the spirit Mercurius, who was strongly associated with alchemy. In his extended investigation into the psychological meaning of alchemy, Jung identified Mercurius with Hermes and listed some of their varied and paradoxical attributes:

> Mercurius, it is generally affirmed, is the arcanum, the prima materia, the "father of metals," the primeval chaos, the earth of paradise, the "material upon which nature worked a little, but nevertheless left imperfect." He is also the ultima materia, the goal of his own trans-formation, the stone, the tincture, the philosophic gold, the carbuncle, the philosophic man, the second Adam, the analogue of Christ, the king, the light of lights, the deus terrestris, indeed the divinity itself or its perfect counterpart.[139]

Some of the items in this list—the prima materia, the stone, the tincture, for example—are purely alchemical terms which refer to the raw material with which alchemists worked. A number of these terms, however, are rather elevated: Mercurius is said to be an an-alogue to Christ, or God himself, or is the perfect counterpart to God. Yet Jung notes that Hermes/Mercurius is also associated with lasciviousness and obscene pictures of the alchemical marriage—a symbol in alchemy of wholeness, of the goal of the alchemical work—which, Jung asserts, were preserved as pornography. Hermes is also present in pictures of "excretory acts, including vomiting." These belong, says Jung, to the "sphere of the 'underworldly Hermes.'" This contrast between what is elevated and what is base is central to understanding Hermes and the trickster god and his role in relation to synchronicity, for the Trickster is nothing if not paradoxical. The Trickster contains but is not contained. He can bob up anywhere because he makes the world in which he moves. This uncontrolled and unpredictable aspect of Hermes means that synchronistic coincidences—bringing, for instance, good or bad luck—can crop up anywhere.

THE TRICKSTER IN THE MARKETPLACE

Hermes' ability to reach through the cracks in ordinary reality and make connections between the known and the unknown arises in part, surprisingly, from his ancient role as patron of commerce. In *Hermes the Thief*, classicist and student of Freud, Norman O. Brown, notes that in primitive Greek villages the marketplaces lay at the borders between settlements. "Primitive trade on the boundary was deeply impregnated with magical notions."[140] These notions expressed the mystery of exchange between a village and its unknown and usually feared neighbors. Hermes was the god associated with the exchanges made at these boundaries. Such exchanges were understood to be magical communications between the known of one's own village and the unknown of an alien village. In places sacred to Hermes, a person who wished to trade would leave an object. The exchange would occur silently, the traders never actually meeting. People returned later to find the goods for which they had traded. Even later during classical times the association of Hermes with the marketplace remained as a remnant of this most primitive form of trade.

Hermes' place in ancient commerce illustrates his role in connecting the known with the unknown across borders. In larger terms it demonstrates that his sphere of operation is not bounded by the known but encompasses "the totality of existence." Consequently, Hermes the Trickster can be found symbolically wherever rigid notions of life exclude part of life's totality. He can appear at any boundary, including those we ourselves set, even unconsciously, or those set by our culture.

The world of modern mechanistic science is a world bounded by the rigid constraints of causality. It is the Trickster's predilection to cross such boundaries, bringing the unexpected to the commonplace.

His gift of synchronicity, however, seems dark, sinister, and threatening to that world, because it appears to be an intrusion from an alien landscape, a world that mechanistic science cannot enter. Synchronicity plays the devil with the myth of causality. The expressions of the Trickster, who returns to us the life that our boundary-making tries to exclude, raises a satanic specter in the eyes of science. Its qualities are the most offensive: it cannot be objectively tested, and it makes itself unavailable for prediction and control. Synchronicity represents a hostile other because it is acausal, and as such blasphemes against the mythos of the causality principle.

SLIPS OF THE TONGUE

Freud's notion of *parapraxes* refers to the false but embarrassing acts commonly called Freudian slips. These often reveal hidden truths about our makeup, showing the human bumbler behind the smooth and ordered facade we project to others. A student, for instance, was once sitting through a long and dry lecture when the classroom door blew open, letting in a cold draft. He exclaimed, "Someone shut the bore!" We have all made such slips and can remember them with varying degrees of embarrassment or humor. If we understand the Freudian slip in Jung's terms, however, we are confronted again with the Trickster who, manifesting his power through mistakes and slips, undercuts the boundaries we set for ourselves. When these boundaries fail, a gap opens between who we are, on the one hand, and who we are trying to be, on the other. Through this gap slip the little accidents or mistakes which are the expressions of the Trickster.

We can easily understand how our missteps result from our own imperfections, from pride, from a distorted or inaccurate sense of our selves. But synchronistic events, in contrast, do not at first nod seem to spring as clearly from our personal lives. Their meanings are rarely apparent. To understand the connection between parapraxes and synchronicity, it is useful to examine the former in more detail. Let us begin by looking at the way meaning arises from classical parapraxes.

One of the authors once heard an aging professor give a highly creative talk on the meaning of Ingmar Bergman's film, *Wild Strawberries*. The audience was enthralled. In the excitement of the applause that followed, he stepped forward and exclaimed, "See! As I grow older my *procreative* powers—excuse me!—*creative* powers, grow greater and greater" [emphasis added]. A slip of the tongue may reveal its meaning in terms of unconscious psychological or emotional needs. This slip, which without knowledge of psychology we would take simply as nonsensical, is not difficult to understand in terms of the speaker's fear of the loss of sexual potency as he approaches old age. Such slips unveil a connection between the world of our day-to-day identity and the often hidden world of emotions. Like parapraxes, synchronistic events also connect different worlds: the world of everyday reality with the world of the mythic realms of the unconscious.

Jung's notion of synchronicity is based on the idea that when the unconscious mind is stimulated powerfully in one direction there is a corresponding lowering of mental energy elsewhere. This frees up the psychoid, or undifferentiated level of the mind. This primitive level is free of the ordinary restraints which separate the mind from connections with the external world. When such connections are made, they are made in the form of synchronistic events. As is the case with Freudian slips, a strong emotion or a powerful surge of mental energy can make possible the synchronistic event. Let us consider an instance of each.

One famous clinical case of Freud's involved a well-to-do young woman who was suffering from hysterical blindness—that is, her blindness was due to a neurosis and not a neurological disorder. All that could be established regarding the events surrounding the onset of the blindness was that they involved the death of her father in a Protestant hospital in Italy. Something was wrong with the story, however, and she could not supply the needed answer. That is, not until she slipped one day and referred to the nurses as "prostitutes" instead of Protestants. She corrected herself immediately, but the cat was already out of the bag! Her father had, in fact, died in a brothel, in the arms of a prostitute, and she had been forced to come to the brothel to identify the body. The emotional aftermath of this incident caused her to repress the whole affair and produced her blindness. The emotional charge she was left with, however, opened the gap for the slip of the tongue and provided Freud with a badly needed clue about the real cause of the neurosis.

Marie-Louise von Franz, speaking about synchronicity, tells the story of a man who, suffering from a psychotic idea that he was the savior of the world, attacked his wife with an axe to "exorcise the devil out of her." She called for help, and at the instant a policeman and psychiatrist entered the house, the single electric lamp that lit up the passage where they all stood exploded. They found themselves in darkness covered with fragments of broken glass. The man immediately declared, "See! This is like it was at the crucifixion of Christ; the sun has eclipsed." He took this as a confirmation that he was the savior. Von Franz, however, points out that a light bulb is not the sun—that luminous source of higher consciousness. It is rather a small light made by man, symbolizing his ego. From this view the real meaning of the event is evident. The explosion of the bulb reflected the collapse of the man's ego. The event in this case is speaking the language of dreams, showing the man—could he only see it—exactly what he was blind to in his psychotic state. Needless to say, his mental state was highly energized at the moment of the incident. [141]

While concentrations of psychic energy do not in the usual sense cause synchronistic events, in Jung's model they do allow them to occur. The meaningfulness which gives a synchronistic event its special character arises from the symbolic connectedness between elements in the deep layers of the psyche and some situation in the external world. The synchronistic event is independent of the mental situation which allows it, because the connection between an emotional surge in the psyche and the situation outside is caused neither by the internal nor the external arrangement of things. Rather, the outside event and the inner psychological one—the emotion, the complex, the concentration of energy—unfold simultaneously. The two events, one inner and one outer, connect not by virtue of one causing the other, but by a mutual reflection of a common meaning. Such a connection recalls the medieval notions of sympathetic resonance. It also recalls Bohm's notion that the superimplicate order unfolds patterns of meaning into the implicate and ultimately the explicate orders of reality. In Chapter 2, we suggested that such patterns may take on a variety of forms simultaneously, some physical and some mental. We will return to this notion again in Chapter 6.

When an individual's inner life corresponds in a symbolic way to the outer, objective world, the two are connected by meaning. And since this connection is acausal, it appears accidental, an unexpected interruption or intrusion breaking into the normal order of things. Here, precisely, is the Trickster, revealing meaning in his typically unorthodox way.

THE TRICKSTER IN AFRICA

Through synchronistic coincidences the Trickster can sometimes confront us with what we do not know about ourselves but must recognize if we are to know the whole of our own reality. Canadian priest and student of African culture Robert Pelton, in *The Trickster in West Africa*, refers to the trickster god of the Ashanti people, Ananse, as "pure synchronicity." Ananse introduces into human culture ways of seeing that have remained hidden or left out. The Trickster, however, works in his own distinct way. Pelton notes that Ananse, like Coyote or Hermes, "does not act from noble motives or by a rational, straightforward method, but by faithfully following his own way of making connections—by shattering the accepted boundaries of language, action, and even modes of being."[142] As a creature without boundaries, his existence and activity are never absolutely fixed in place. He makes connections across the limits we ordinarily set for life, bringing together polar opposites and disclosing that which is hidden.

Pelton tells a tale of another West African trickster, Eshu, who plays a prank on two farmers. His joke reveals an important inadequacy in the relationship between these friends:

> Once two friends owned adjoining farms. They dressed alike and were in all ways a model of friendship. Eshu decided to make them differ. He used to walk each morning on the path between the two farms, and one day set out wearing a multi-colored cap . . . and let his staff hang over his back instead of his chest. He greeted the friends, already working in their fields, and passed on. Later they began to argue about the color of his cap and which way he was going. [In one version of the story Eshu returned at this point, walking the other way. This led the farmers to an animated argument, each insisting that the other had been right the first time.] Soon they came to blows. When they were brought before the king, Eshu confessed to igniting the quarrel because "sowing dissension is my great delight." When the king tried to bind Eshu, he fled, started a fire in the bush, hurled burning grass on the town, and then mixed up the possessions that the townsfolk hauled out of their houses. A dreadful row began, and as Eshu ran off laughing, he boasted that everyone had played his game well.[143]

Pelton notes that in this story, the Trickster points up the animosity which underlies the idealized friendship of the two farmers. Always acting from a stereotyped model of friendship, they come to blows when the Trickster walks the boundary that symbolically separates their identities. So in spite of their shared, if inaccurate, perceptions, they disagree about the Trickster because "neither of them can encompass the other's vision. They notice Eshu's clothing, his staff, and his pipe, but neither really *sees* his movement—or what the other sees." Further, Eshu brings the dispute into the society, where even the king cannot control the Trickster as he mixes up the boundaries between the villagers' possessions. The dissension the Trickster sows brings into focus the boundaries between people and the price that is to be paid for repressing or ignoring these boundaries in the name of idealized friendship or good will. As we know, even best friends do not always get along. Apparently the two farmers, and by extension the society, had forgotten their real underlying differences. This state of affairs attracted the notice of Eshu, whose game, after all is said and done, restores to the culture a healthier, more realistic attitude toward human relations.

The trickster in this story accomplishes his effect symbolically—by wearing backward the symbols of manhood, the staff and the pipe, and by burning the people's houses and mixing up their possessions, both of which represent their individual identities. His effect appears chaotic, but that is only because it compensates for the repressive imbalance of too much order, as symbolized in the story by the idealized friendship of the two farmers.

The Trickster first appears on the borderline or threshold between two individual identities. He belongs, however, outside the normal social order, as his disruption of the king's attempt to settle the dispute shows. His trickery comes as a visitation from outside, though from a psychological point of view it was invited, so to speak, by the internal psychology of the two farmers. There is a paradox in the Trickster's nature: he can encompass opposites by bringing together both the conscious and the unconscious characteristics of the farmers and the villagers. The story also shows the Trickster involved in a further paradox: it shows him making the world as it already is, not as the two farmers have idealized it. And, tricksterlike, he accomplishes this feat for his own selfish ends.

THE TRICKSTER IN THE MODERN WORLD

In our own culture, the notion that we can understand our lives by tracing all events and situations to understandable causes has become a powerful obstacle to confronting relationships unconnected by chains of cause and effect. Thus, we are often oblivious to meaningful coincidences, even when they are striking—thinking them no more than our share of chance. Statisticians encourage this view; however, this attitude is a holdover from the Newtonian mythos in which people are seen as analogous to atoms that are knocked about by random Brownian motion, forming brief, illusive, and seemingly intelligent patterns, but at root amounting to nothing but dust devils of perception. This attitude throws up a wall against the irrational part of life, closing us off from symbolic meaning.

Few would suggest that significant coincidences never occur by chance alone—not even Carl Jung himself. Sri Aurobindo, the great Indian yogi and sage, for instance, once wrote:

> The figure of the world reveals the signs
> Of a tied Chance repeating her old steps
> In circles around Matter's binding posts.
> A random series of inept events
> To which reason lends illusive sense, is here.[144]

But Sri Aurobindo did not mean that all or even most coincidences are "inept events to which reason lends illusive sense." To come to such a conclusion would be like seeing a man picking up coins from the street and assuming that all of his money was acquired in this way.

Perhaps we also tend to overlook synchronicity because it is a symbolic form of expression, and we simply do not understand it. Symbols are, at best, difficult to understand because their meaning is open to varied interpretations and can never be completely grasped. Unlike allegories, which have fixed and entirely comprehensible meanings, synchronistic events are symbolic; they are not in the end reducible to any form of expression other than that which they originally take. We can gain insight into the symbolic meaning of a particular synchronistic event or dream, but it will still retain something of the mystery that belongs to the truly numinous. Such numinosity indicates the source of both symbols and synchronicity in the depths of the chthonic underworld, from and to which Hermes is the messenger.

Because symbols and symbolic events are rooted in this under-world, even the most careful and painstaking examination can never reduce their mystery—the mystery of life—to a rational equation. This very fact gives synchronicity both its irritating and its enchanting qualities, qualities reflected in the characteristics of that bringer of both good luck and bad, the Trickster. The symbolic meaning of a synchronistic event arises from the correspondence between the deep-est layers of the human mind, about which we really know very little, and an external world in which final causes remain, at best, a mystery to us. A synchronistic event points to the correspondence between two deep mysteries, exposing a connection between them in symbolic form.

We live in a world in which, for the last fifty or sixty years, subatomic physics has described a universe founded at bottom on acausal connections, on paradoxical and seemingly illogical rela-tionships and observations. Yet as a culture we still deny acausal, symbolic connections as part of our lives and the lives of our souls. As a result, the Trickster continues to play the devil with us, con-tinually upsetting with his sudden windfalls (the ancient Greeks re-ferred to the offerings left at roadside herms as *hermaion* or windfall) the rigid notion that everything is explicable in terms of cause and effect, that everything can be understood by the rational faculty.

The Trickster puts life in our path in spite of our denials. We continue to stumble over his gifts, ignoring their disturbing nature when our luck is good, cursing some vague fate when our luck is bad. As Pelton observes, the Trickster

> Enters the human world to make things happen . . . to break and reestablish relationships, to reawaken consciousness of the presence and the creative power of both the sacred Center and the formless Outside. Then he returns to that hidden threshold which he embodies and makes available as a passage "to save the people from ruin."[145]

Hermes, Ananse, Coyote, or tricksters by any name bring the divine world, both within and without, into human life by passing through the hidden boundaries which we create and yet remain unaware of. Through the Trickster's play we enter myth, the imaginative awareness of life that answers questions about who we are, where we have come from, and where we are going. This entrance we make symbolically, through acausal connections, the synchronicity that is the gift of the Trickster.

6

THE MEANING
OF SYNCHRONICITY

*If you stand right fronting and face to face to
a fact, you will see the sun glimmer on both
its surfaces, as if it were a [scimitar], and feel
its sweet edge dividing you through the heart
and marrow, and so you will happily conclude
your mortal career. Be it life or death, we crave
only reality.*

HENRY THOREAU
Walden

Lighten up!

TOM ROBBINS
Jitterbug Perfume

Synchronistic coincidences are al-
ways personal events. The fascination of synchronicity is to be found
in the feeling that each coincidence speaks to us individually. As we
have seen, however, the meaning of the message is often far from
obvious.

Perhaps because synchronistic coincidences are objective events
that symbolically seem to mirror personal, subjective realities, they
are often difficult to comprehend. To understand them we must seek
a knowledge that binds together the objective world of physical events
with the inner, personal world of subjective reality. David Bohm's
notion of the implicate order does this to a degree. But his theory is
new and still inarticulate in dealing with personal meaning. Carl
Jung's thought is strong just where Bohm's is weak. Jung's ideas

provide a rich understanding of inner, personal, and psychological meaning, especially in the realm of archetypal patterns, while at the deepest psychoid level they participate in the *unus mundus*—one world beyond all separations into subjective and objective realities. At root, archetypes touch the *unus mundus*, carrying into the psychological life of each individual the mythological motifs that characterize them.

One archetype that we have already looked at in some detail is of particular importance to synchronicity: the archetype of the Trickster. He is the mythic embodiment of the unexpected. He symbolizes the unexpected eruption into awareness of truths hidden away from the ego. In a psychological sense, the Trickster is one mode by which other archetypes, such as the archetype of the self, assert themselves. For instance, Hermes was both a messenger to the gods (archetypes) and herald to them. He was always depicted carrying the herald's staff. Thus he brings news of the gods' wishes, and he also heralds their coming. On a psychological level synchronicity plays both of these roles for the archetypes. Synchronicity can signify the unconscious activity of an archetype and herald its arrival as a powerful influence in an individual's consciousness.

After completing his essay on synchronicity, Jung saw a vision of the Trickster on a wall of his Bollingen house, a vision he later carved in stone. This appearance of the Trickster is characteristic of his style: he pops up unexpectedly. The quality that he brings to synchronicity, however, is not simply that of surprise. His manner has the impish charm of cunning and magic. There is a flavor of roguish enchantment to the situations he orchestrates. American Indian tricksters were frequently known as "delight makers" in honor of this winsome deviltry.

This mysterious side is seen in Hermes, who was often associated with the night and the sense of magic that can accompany the night. This sense, however, is not limited to the night alone but can also appear by day "as a sudden darkening or an enigmatic smile. This mystery of night seen by day, this magic darkness in the bright sunlight, is the realm of Hermes."[146]

One has the feeling that in synchronicity the Trickster engages in the fabulous play of a divine jester; he is a "juggler of reality."[147] It is in the notion of play, we believe, that we may find the key to understanding our best relationship to the Trickster and thus to synchronicity. It is also the key to discovering his divinity in ourselves.

In this concluding chapter, we will see that the Trickster is capable of as many varieties of play as there are varieties of synchronicity. As a messenger and herald he represents the interests of a considerable range of unconscious or mythic figures. The most roguish play of the Trickster, however, is in the role of the prankster. Here he sticks us with one after another of those annoying, stumblebum coincidences known in German as "monkey tricks." In these the Trickster acts on behalf of an unconscious structure known as the *shadow*.

SHADOW PLAY

The shadow is Jung's term for the flip side of the outward personality that each of us has crafted for the benefit of others as well as for ourselves. Jung termed this outward personality the *persona*, the name used for the masks worn by actors in ancient Greek dramas. Actors did not communicate their moods directly by their facial expressions, as is done in the theater today, but rather by wearing masks. The shadow, on the other hand, is made up of those personal qualities we most vigorously resist admitting about ourselves. Ironically, these are often the very qualities that we intensely dislike in others. Try bringing to mind, for example, a person whom you dislike with particular intensity. If you can provide a description of that individual, you almost certainly may read in it the characteristics of your own shadow. The strong emotion you feel toward the other person indicates the presence of the shadow, in projection, as those traits which you have repressed just for what you believe are the best of reasons. In more technical language, the shadow consists of those qualities in ourselves which we repress or at least deny most of the time for the sake of the *ego-ideal*, all those learned values, both conscious and unconscious, which we have been raised to accept as forming a proper guide to our behavior and our identities.[148]

Normally, one thinks of the shadow as embodying the most undesirable features of the personality, and this is usually the case. Indeed, the most positive person can sometimes hide the most negative shadow. A charming individual may hide an extremely critical side as well as keeping a bad temper under wraps. On the other hand, it is possible to have a shadow that is more positive than the persona. This situation occurs when someone actually identifies with the undesirable characteristics in his own personality. Part of the appeal of Humphrey Bogart, for instance, was his tender, caring shadow, all too visible under his tough-guy persona.

The personal shadow in each of us, which we have been discussing so far and which is formed from our own personal history and psychological makeup, is part of a larger mythic or archetypal shadow, which belongs to the unconscious. When we project our personal shadow onto others—seeing in them our own concealed nature—or when we likewise project it onto the world of inanimate objects, we reduce our awareness of ourselves to the degree that the personal shadow may become enlivened by its archetypal likeness. It then acquires a life of its own, often taking on the aspect of the prankster and playing tricks on us.

Since the personal shadow, in fact, is nothing more than an odds-and-ends collection of traits that we have failed to fuse into our conscious persona, the tricks played by this prankster are likewise disorganized. We find ourselves experiencing a day in which, to use Jung's words, "everything goes wrong and nothing intelligent happens except by chance."[149] We are beset by one trifling difficulty after another. The sheer number that can be packed into the first few hours of a day is astounding! We say we are having "one of those days." At such times all phone calls connect to busy lines, the car needs gas, the weather is terrible and the umbrella is at the office, and the traffic lights are all red. It is just this variety of bad luck that evidently inspired Murphy's Law, which states that anything that can go wrong will go wrong. There are numerous correlates; for instance, if a slice of bread falls on the floor, it will always land butter-side down. One of the authors likes to refer to such coincidences as "perverse synchronicity."

What can be done at times like this? Marie-Louise von Franz observes that at such times one is, in a sense, outside of oneself. In most cases the solution, then, is simply to relax and get back in touch. "Monkey tricks" typically happen when one is in a hurry and "beside oneself." The key is to relax into the moment, realize that life will not end if the next stoplight turns red, and try to be centered on yourself. Doing this can lead to surprising results. You may find, for instance, that the tricks are actually not so bad. Upon regaining your composure, you often discover that each trick is followed in time by another coincidence that provides a solution to it. How many times, for example, have you had something go wrong with a car, then discovered with relief that you were remarkably near a garage where it could be fixed?

In the service of the shadow, however, the prankster can be more than a petty nuisance. If a major part of the personality is barred from consciousness so that it is banished to the shadow, the consequences can be serious psychological problems as well as abundant synchronicity of the prankster variety. A case in point was Jung's own friend Wolfgang Pauli who, it turns out, was strongly disposed to the intellectual side of his personality much to the expense of his feelings. The latter were often relegated to his shadow:

> In Pauli's case, thought had dominated feeling so that the emotions were relegated to what Jung termed the Shadow. . . . Thought, sensing what it felt to be primitive forces at work, put the lid on even tighter so that Feeling found itself in the position of a red-hot pressure cooker with the valve jammed.[150]

What resulted was an utterly disrupted personal life, vehement and unnecessary sarcastic attacks on fellow scientists, and frequent bouts of drunkenness. What also resulted was the famous Pauli effect, by which delicate laboratory instrumentation collapsed upon Pauli's mere appearance!

Hermes the Thief

The Trickster's play frequently gives us opportunities, usually unwelcome, for personal growth by flaunting our most private secrets for the whole world to see. This seems to be the Trickster's delight. Wearing the cloak of the shadow itself, he exposes its secrets through synchronistic coincidences, putting them out in public where we meet them head-on. Thus, the play of the Trickster makes us confront our own faults in the everyday world, much as we are forced to confront them in our dreams. These instances offer the opportunity to recognize our faults and, by owning them, to take away their sting and in the bargain render ourselves more whole.

Like tricksters in many cultures, Hermes is the archetypal thief, stealing our purposes and turning them to the service of his own trickery. Just when we want to make our best impression, for example, we make some silly mistake, mispronounce a name, spill the coffee, do little embarrassing things. This is the Trickster as the shadow, stealing our purpose when we want to appear flawless—just to amuse himself with our foolishness. If we are open to this impish play, we realize that we have been reminded that we are only human, that we have limitations, no matter how perfect we might wish to appear.

Interestingly, theft as a means to wholeness is not unknown, even on the social level. According to Norman O. Brown, in some cultures thievery is an accepted form of exchanging goods and is distinguished from robbery or burglary.[151] In such cultures it is the custom to take certain objects that one likes from another's home. It is as if you were visiting a friend and saw an attractive vase on the mantelpiece. You could simply take the vase home with you—an act of thievery that would be acceptable to everyone in the community. The practice reinforces the bonds between members of the community and serves to strengthen everyone's sense of belonging. A similar legitimate form of theft can also be found in one of the ancient Greek festivals of

Hermes, on Samos, at the feast of Hermes Charidotes, "where the populace was allowed to steal and to commit highway robbery."[152] In these customs we see that a selfish act can create some good for everyone: the desire to possess something for oneself becomes a way of strengthening the communal bond through exchange and by reminding people that distinctions based on property are always in some sense artificial, beside the point in the context of everyone's shared humanity and mortality.

Typically, we miss the opportunities for greater psychological wholeness provided synchronistically by the trickster-as-shadow because we are too busy protecting ourselves by projecting the problem outward onto others or onto a hostile fate. When this happens, the Trickster may put the problem before us again and again, and despite our mounting frustration force us finally to deal with it. In such cases, though the Trickster acts for himself, as it were, we can detect behind him the directing power of the archetype of the self, which would move the psyche as a whole to greater maturity.

One persistent encounter with the shadow is colorfully portrayed in the Persian tale of Abu Kasem's slippers.[153] These slippers were famous in all of Baghdad, as famous as their tightfisted owner, Abu Kasem himself. For though Abu Kasem was wealthy, he was a miser, and the extraordinarily shabby state of his slippers portrayed his own nature all too well. The poorest beggar in town would have been ashamed to be seen wearing them!

One day Abu Kasem completed an unusually lucrative business deal by which he obtained a quantity of fine attar of roses from a bankrupt perfume merchant, as well as a set of choice crystal bottles to decant it in. To celebrate, he treated himself to a rare visit to the public baths. There his disgusting slippers were so conspicuous that an irate fellow merchant lectured him on the disgrace they brought to his name. Abu Kasem, however, replied that the slippers were really not so badly worn that he could not use them. Later, while he was in the bath, an illustrious judge arrived and left his clothes near Kasem's garments as he entered the baths. Upon leaving, Abu

Kasem discovered an excellent new pair of slippers in the very place he had left his own—or at least in almost the very place. Evidently, his irate companion had done him a good turn and replaced the old ones for him. He slipped on the fine new slippers and left. The judge, however, upon preparing to leave the baths, was beside himself. His slaves looked everywhere for his slippers but located only the disgusting pair, which everyone knew belonged to Abu Kasem.

The judge wasted no time hauling Abu Kasem into court and, knowing that he was rich, fined him heavily. This apparently was enough for Abu Kasem. In a fit of temper he decided to get rid of the slippers, hurling them out the window where they fell with a splash into the river Tigris, which flowed muddily by his house. A few days later they tangled and ripped the net of local fishermen who, recognizing the revolting things, flung them back through Kasem's window where they landed with a crash on the table, breaking the crystal bottles to pieces and spilling the perfume out into a soggy mess mixed with the mud from the slippers. Abu Kasem was enraged.

Determined to rid himself of the slippers, he carried them to his garden where he proceeded to bury them in the ground. A neighbor, however, observed him digging and concluded that he had treasure buried there. As the Earth and everything in it belongs by law to the caliph, the neighbor, being no friend of Abu Kasem, took himself off to the governor's palace to report the matter. Abu Kasem was quickly called before the governor to account for himself, but no one would believe his preposterous story about the slippers! He was again given a heavy fine.

Desperate to rid himself of the accursed things, Abu Kasem carried the slippers out into the country and threw them into a pond, watching them sink toward the bottom. Unfortunately for him, however, the pond was part of the water supply of Baghdad, and the slippers swirled to the mouth of the pipe and stopped it up. The city workers whose job it was to unclog the pipe instantly recognized the slippers and reported Abu Kasem to the governor for befouling the city's water. Abu Kasem was again forced to pay a large fine.

Now, very nearly broke and in a state of despair, Abu Kasem decided to burn the damned things. He set them out on his balcony to dry, but a dog snatched one of them and, playing with it, dropped it down to the street below where it struck a pregnant woman on the head, causing her to miscarry. Her husband ran to the judge to demand recompense. Abu Kasem, now a broken man, was forced to pay. Before leaving the court, however, he raised the terrible slippers and cried: "My lord, these slippers are the fateful cause of all my sufferings. These cursed things have reduced me to beggary. Deign to command that I shall never again be held responsible for the evils they will most certainly continue to bring upon my head." It was said the judge could not reject this plea, and so ends the story.

What does this tale mean? It can be read on a variety of levels, but our interest is in the Trickster and the shadow. The slippers are, if nothing else, an outward sign of Abu Kasem's own miserliness, a miserliness that he himself is prone to deny, though it is all too apparent to the rest of the world. Even when confronted directly in the public baths about the slippers' dreadful condition he attempts to minimize it, saying that they are not so bad that he cannot use them. He will not admit to this important, indeed central, part of his own character; he has relegated his closefistedness to the shadow. As a result of this denial, the Trickster steals Abu Kasem's purpose and turns it to his own ends. Thus is Abu Kasem confronted with his own shadow, and an opportunity for greater insight into his own real nature.

We must realize that the shadow is not made up entirely of deeply repressed material; much of it consists of personal qualities we simply will not face up to in ourselves. Miserliness can be such a quality. We rarely admit our own stinginess, even to ourselves, and when it becomes an obsession we admit it even less. Abu Kasem was a miser, but he was not forced to confront his own closefisted nature until the incident at the public baths when his companion lectured him on the state of his slippers, that is, upon his miserliness. His loathing to admit this basic part of himself is the opening for the Trickster to steal Abu Kasem's own defense and turn it from a denial of the shadow into the series of troubles of the sort the Trickster enjoys. He does so first of all in a fashion reminiscent of the West African trickster, by mixing up Abu Kasem's personal items with those of the great judge. Had Abu Kasem been wearing slippers appropriate to his own wealth the problem would have been minor, but this was not the case. The result was the first of many terrible fines that he would have to pay.

The severe consequences of the event at the baths should have been enough to cause Abu Kasem to admit his miserly nature, but his habit of denial was deep, so instead he tried to toss it off, literally, by throwing the slippers out the window. This did not work, as the synchronistic Trickster intervened with dire effect.

Now Abu Kasem gets serious about covering up his flaw. He first tries to bury the slippers in the ground and later to bury them under the water at the bottom of a pond, efforts which suggest that he is attempting to shove the whole business back down into the unconscious. None of this works, of course. The Trickster will not allow it.

The final effort to rid himself of the slippers by burning them seems more authentic than the previous ones. The metaphor of burning away one's obsessions is a common one, but Abu Kasem's habit is too strong and even this does not work for him. Nevertheless, we see that after this first sincere effort, the judge takes some pity on him, agreeing not to hold him responsible for the further pranks of the slippers. Following the effort to burn them, this decision seems much like that of a modern judge who resolves to overlook certain offenses if the accused is making a serious effort to change.

Stolen Intentions

When the Trickster steals away one's purpose and turns it to his own amusement, he brings the hidden shadow out into the daylight in the bargain. Things turn out quite differently than we planned. Since the Trickster is often connected with storytelling, and in the case of

Hermes even with writing, we will illustrate this point with an example from a well-known novel, *The Adventures of Huckleberry Finn* by Mark Twain.

The story is about Huck's trip down the Mississippi River with a runaway slave named Jim. Huck is a white boy who is part rogue and part rebel. Jim, in contrast, is a black man whose ways are foreign to the middle-class society into which Huck has been temporarily adopted by the Widow Douglas. He comes from outside the white culture, which includes all the persons Huck most looks up to—people like the widow or Tom Sawyer, who perhaps most clearly represents Huck's ideal. As a black man and a slave, Jim is outside everything that forms Huck's ideals of who he is and who he should aspire to be. What is more, as literary critic Daniel Hoffman points out, Jim is knowledgeable in his own superstitious way of both the mysteries of nature and the spirits who people an unseen world surrounding both blacks and whites, and so Jim is a link to other realities.[154]

Jim in his own right is very human, and at times quite touching. His humanity is not always apparent to Huck, who in moments of dissension is prone to dismiss him as simply an "ignorant nigger." In typical trickster fashion, though, Jim's humanity is brought home to Huck when a joke he tries to play on Jim backfires. This occurs at a point early in the journey down river when Huck and Jim become separated, lost in a thick bank of fog. Huck is in a canoe, and Jim remains on their raft. They spend most of the night on a dangerous stretch of the river, calling to each other but unable to make contact. After several hours each falls asleep exhausted, but before dawn Huck sights the raft on the now-clear river and paddles over to it, where he finds Jim asleep amid a litter of mud, leaves, and broken branches deposited on the raft when it ran into the bank of a small island during the night.

When Huck climbs up onto the raft, he decides to play a trick on his sleeping companion. He lies down, then feigning noises of awakening, he rouses Jim from his sleep. When Jim sees Huck he is overjoyed. He was sure that they had been separated permanently and was afraid that Huck had drowned during the night. But Huck, thinking to play a trick, feigns astonishment, asking Jim if he is drunk. He succeeds in convincing Jim that everything he is talking about must have happened to him in a dream. Always ready to credit anything to the supernatural, Jim accepts that he must have had quite a dream, and begins right away to interpret it. Huck allows him to go on making a fool of himself until the dawning light of the morning reveals the broken branches and scattered leaves that cover the raft. Just as Jim finishes his interpretation of the dream, Huck points to the litter and asks, "What does *these* things stand for?" As far as Huck is concerned, this question is the punchline of his joke and the end of the whole affair.

But the Trickster steals Huck's joke from him. Jim turns to look around the raft; slowly he realizes that he's had no dream at all and that Huck has made a fool of him. But he is in no mood for joking. He'd thought during the night that he'd lost his only companion, a boy he'd come to love, and so when he sees that Huck has tried to humiliate him, Jim turns to him and says:

> What do dey stan' for? I's gwyne to tell you. When I got all wore out wid work, en wid de callin' for you, en went to sleep, my heart wuz mos' broke bekase you wuz los', en I didn't k'yer no mo' what become er me en de raf. En when I wake up en fine you back agin', all safe en soun', de tears come en I could a got down on my knees en kiss' you' foot I's so thankful. En all you wuz thinkin 'bout wuz how you could make a fool uv ole Jim wid a lie. Dat truck dah is *trash*; en trash is what people is dat puts dirt on de head er dey fren's en makes 'em ashamed.[155]

After delivering this speech, Jim walks into the raft's little wigwam, leaving Huck feeling ashamed enough to tell us, "I could almost kissed *his* foot to get him to take it back." After fifteen minutes of struggle with himself, he goes to Jim and apologizes, telling us "I warn't ever sorry for it afterwards, neither. I didn't do him no more mean tricks, and I wouldn't done that one if I'd a knowed it would make him feel that way." Here, the reversal of Huck's joke becomes a revelation of Jim's humanity, because Huck's own sentiments force him to realize to his shame the other's feelings, to realize how he has hurt Jim.

On the psychological level, the Trickster, who capers for his own amusement, is not responsible for Huck's change in attitude toward Jim but only for the situation which provides Huck with the opportunity for change. That Huck was able to take advantage of it is a measure of his own humanity.

While the archetypal trickster Hermes was both a thief and a patron of thieves, we may also recall that, ironically, he was at the same time a patron of travelers. In the episode above we see Huck's joke stolen and turned against him. Huck is thus given a humbling lesson in understanding his friend and companion as a fellow human being. To appreciate the Trickster fully, however, we must not stop with his antics as a thief. We must turn to his complementary role as patron of travelers.

Traveling implies a goal—a movement from one place to another—and so it can represent, as it often does in mythology and literature, a journey toward growth. We see this in novels of the picaresque tradition such as *Tom Jones* or *The Adventures of Huckleberry Finn*, which usually trace the progress of a central character through several episodes that expose the follies of whatever society he happens to belong to. Often he grows into maturity through a series of rough or humorous adventures, satire, comic reversals, and surprises which signify the presence of Trickster.

The preceding episode from *Huckleberry Finn* also points up a more subtle side to the Trickster, a side which comes from Hermes' role as guide of souls. The ancient Greeks believed that after death the soul, or *psyche*, was guided to the underworld by Hermes, who could also guide persons out of the underworld, as he guided Eurydice when she began to follow Orpheus from Hades.[156] This underworld of Hades was referred to as a place of riches, ruled by Pluto, a name which means wealth or riches. In terms of its psychological reality, therefore, the underworld of the unconscious is a place of riches for the psyche. Hermes guides the psyche to the unconscious as, for example, in dreams and in so doing is the patron of travelers into the richness of soul. He is also the god who brings the contents of this underworld out in the open by his synchronistic tricks. When Huck's joke on Jim is turned against him and that reversal helps Huck see into Jim's real humanity, Huck is being led into the richness of the soul, the real life that is in Jim as a human being. Here Huck is guided toward the source of life.

As Trickster, Hermes acts for his own delight yet gives us the opportunity for insight into our own darkness. Look for the archetypal trickster in any story of growth, for growth always means moving toward one's own human richness which, in turn, means the growth of one's soul.

SYMBOLIC PLAY

The Trickster delights in frolicking with symbols. Synchronicity in such instances seems to mirror processes that are afoot deep in the psyche. Jumping across boundaries between the conscious and the unconscious, between the psychological and the physical, the Trickster tosses out images in play that express the sheer vitality of the imagination. An example, described in Chapter 4, is the appearance of the series of fish that Jung encountered while writing on the meaning of the fish symbol. The appearance of the fish literally mirrored Jung's current passion. If we take his own analysis seriously, however, it means a good deal more: fish symbolize "the nourishing influence of unconscious contents, which maintain the vitality of consciousness by a continual influx of energy; for consciousness does not produce its energy by itself.[157] In Jungian psychology, the appearance of fish in dreams, fantasies, or synchronistic coincidences usually is taken as a positive sign that the essential life of the unconscious is active. This means that the conscious mind is being "nourished" by the life-giving vitality of the unconscious. Certainly Jung was unusually well nourished.

Another example of the symbolic play of the Trickster, one known personally to us, involved butterflies and a three-year-old boy. At the time, the boy suffered both physically and mentally from a variety of allergies that frequently made his life miserable. The day of September 22, 1986, however, seemed to mark a watershed in his return to health. On several occasions that day the boy was visited by the appearance of butterflies.

He was relaxed and at peace with himself on that day, exhibiting the full vitality of childhood. In the morning he spontaneously painted several large, bright-colored butterflies. On a late-afternoon walk, he and his father saw a number of large butterflies soaring like birds high in the trees. Butterflies appeared again during dinner, which the family ate outside. Late in the evening his mother read two stories to him, both of which, by seeming coincidence, mentioned butterflies. The first concerned fairies that were said to fly about like butterflies. The second was the Chinese tale of the philosopher who dreamed that he was a butterfly, and upon awakening wondered if he could, on the other hand, actually be a butterfly dreaming that he was a person. These were from a book of stories that they were reading through systematically, one or two stories each night.

The butterfly is an ancient symbol, and its meaning is not difficult to guess. Perhaps no living creature goes through such a dramatic transformation and emerges with such beauty as the caterpillar that changes into a butterfly. Hence, the butterfly represents transformation, in this case a healing transformation. The boy in question did not get well immediately, but this day more than any other seemed to be the one on which he began to move toward health.

Careful examination of synchronistic coincidences often reveals that they compensate for personal needs or inadequacies. In Chapter 5, for example, we noted the importance of compensation in the meaning of dreams; here we extend a similar notion to synchronicity.

In some cases compensation is almost obvious. For example, an acquaintance of the authors gave the following account of her father, a previously hardworking businessman who had recently retired. In his free time he had become interested in odds-and-ends repair work around the house: plumbing, a bit of carpentry, minor electrical work. He had not previously been a great handyman, but it was obvious that such work had become a source of satisfaction for him, compensating as it were, for the loss of his previously engaging professional life and easing his transition into retirement. Early in the summer he and his wife spent a week with our acquaintance's family. During that week more things went wrong with their house than during the entire previous six years that they had owned it. Windows jammed, doors stuck, light bulbs flashed and burned out. One evening a pipe under a bathroom sink split open for no apparent reason and began to spray water. The father was busy every minute fixing one thing after another. When he left, the problems abruptly disappeared. It was as if the father's satisfaction from such work had generated all the necessary conditions of needed repairs.

But compensation in synchronistic events is often far from apparent. Take, for instance, the appearance of the beetle at Jung's consulting room window at a crucial moment in the therapy of a woman who was straitjacketed by a neurotic and inflexible view of reality. Jung was her third therapist, and he commented that, "evidently something quite irrational was needed which was beyond my powers to produce."[158] As we know, that something, in the form of the beetle, was supplied by the Trickster. The very irrationality of its appearance, in her eyes at least, compensated so dramatically for her rigid beliefs that it jolted her into the beginning of a new worldview.

When dealing with compensation it is as dangerous to make a literal interpretation of synchronistic incidences as it is to assume such an interpretation of dream events. A case in point is the man, noted in Chapter 5, who experienced the psychotic delusion that he was the savior of the world, attacking his wife with an axe to exorcise the devil out of her.

One does not have to be psychotic to have an inflated self-image. We are all at times prone to think too highly of ourselves. At such moments the Trickster may pay us a visit in the guise of a prankster, to bring us back to earth, to make us look foolish or ridiculous in little ways just when we want to look our best. He is adept at undercutting self-inflation.

Recall Eshu, the West African Trickster who for the sheer delight of it set friends against each other, exposing the flaws in the luster of their perfect relationship. By claiming to be ideal friends rather than imperfect companions with normal human limitations, they invited the Trickster's mischief. The energy that they had put into denying the human part of their friendship returned in the power of Trickster to plague them. Yet despite their strife, they were actually given more life rather than less.

Eshu's pranks compensate for the superficial gloss of the farmers' idealized friendship. The point here is not that they were less than real friends but rather that their relationship, in its superficial perfection, was less than whole, and so they were themselves less than whole. It was Eshu's troublesome prank that provided the opportunity for them to become complete.

Lame Deer, medicine man of the American Lokota Sioux (the western Sioux), tells us of his tribe's clowns or tricksters, who play a similar role in bringing wholeness. Anyone can become a clown through no choice of his own, but simply because he dreams about lightning, the "thunderbirds." When he awakens in the morning he has become a clown, an "upside-down, backward-forward, yes-and-no man, a contrarywise."[159] Now he must perform his dream for others to see; he has no choice:

> If I had a *Heyoka* [clown] dream now which I would have to re-enact, the thunder-being would place something in that dream that I'd be ashamed of. Ashamed to do in public, ashamed to own up to. Something that's going to want me not to perform this act. And that is what's going to torment me. Having had that dream, getting up in the morning, at once I would hear this noise in the ground, just under my feet,

that rumble of thunder. I'd know that before the day ends that thunder will come through and hit me, unless I perform the dream. I'm scared; I hide in the cellar; I cry; I ask for help, but there is no remedy until I have performed this act. Only this can free me. Maybe by doing it, I'll receive some power, but most people would just as soon forget about it.[160]

What sort of performance must one put on that is so embarrassing? What must one act out? Obviously it is not something that one is proud of; rather it is something that one would much rather keep under wraps. We are clearly dealing with the shadow here, and the clown, much to his mortification, is forced to act it out for everyone's enjoyment. This is compensation at its best. He can no longer hide this side of himself but must ultimately join in the laughter himself, accepting his skeleton in the closet as a joke. What's more, he is probably not that much different from everyone else. Others would have found themselves just as mortified had they been in his shoes. Not only does he confront his own hidden side, but others through watching him do the same in laughter. Humor is considered sacred among the Sioux, and the whole affair is not to be taken too seriously. It is the opportunity for a good belly laugh, and everyone is made more whole by it.

Synchronistic coincidences are often compensatory; like dreams, like the antics of the Lokota clown, they place before us the facets of reality that we have previously swept under the rug. In this way the Trickster provides us with the possibility of greater self-understanding, though, unlike the Lokota clown, where synchronicity is involved we are not usually forced to take advantage of it. Huckleberry Finn exemplifies someone who has the grace and humility to learn from the experiences his own trickster puts him through.

INDIVIDUATION AND THE SELF

The turn of the Trickster's play is not always toward the part of the prankster. If shadow dancing brings out the worst in him, if he is at his most roguish in the service of the shadow, then he is at his best in the service of the central aspect of the personality, the archetype Jung called the Self. His appearance in connection with this archetype can mean that synchronistic coincidences are associated with the growth process termed *individuation*. Such growth occurs when the entire psyche comes under the governing influence of the Self, most often in midlife.

Jung described the Self as the archetype whose center is everywhere and whose circumference is nowhere. To feel the pull of this archetype is to feel the pull of destiny, to sense in your most profound conscience your unique station and purpose in life.[161] Jung says that we may think of this deepest voice within us as the voice of God. Such a voice calling us to our destiny usually comes after one has succeeded in achieving the goals of youth. These "heroic" goals, such as climbing the career ladder or gaining the respect and admiration of one's peers, are *collective* goals in the sense that they are shared by all—there is little that is unique in them. Individuation, on the other hand, means to find one's uniqueness, to fully become an individual. It is usually accompanied by a period of disintegration of the old heroic goals, experienced by the ego as a crisis. This is the *midlife crisis*, and it cannot be resolved by the ego alone. There must be intercession by the highest source of purpose within the individual, the archetypal Self.

The Trickster's connection to the archetypal Self is evident in various mythologies in the stories that present him in the role of a mediator between human beings and the highest gods, as Hermes mediates between humankind and Zeus. Hermes' role as messenger of the gods, as well as his reputation as the friendliest of the gods to people, both suggest this mediating function. Pelton recognizes such a role in the West African trickster as well. It fits in naturally with our knowledge that the Trickster is a god of boundaries, especially of the boundaries between the known and the unknown. The highest gods are often distant gods, unknown directly by mortals, and the Trickster is there to bridge that distance.

His presence in liminal or threshold states means, in Pelton's words, that human beings have access through him to a "source of creative power." By playing his tricks and functioning as a guide to the soul, the Trickster "enters the human world to make things happen, to recreate boundaries, to break and reestablish relationships, to reawaken consciousness of the presence and the creative power of both the sacred Center and the formless Outside."[162] In these last few words, Pelton's description corresponds almost word for word with Jung's description of the archetypal Self.

An example of the influence of the Self that involves both a dream and a synchronistic coincidence is described by Robert Johnson in his book, *Inner Work*. It concerns a woman who in a dream found herself in a monastic cloister, sitting with crossed legs in *zazen* style in a cell connected to the chapel, but separated from it by a grille. In the dream she experienced a state of great tranquility as she heard the Mass performed nearby. Though alone in her cell, she closed her eyes and received communion. At the end of the Mass she discovered flowers blooming beside her chamber.

This woman, who had grown up in a Catholic family, had early in adulthood rebelled against her religious background, becoming involved in the practice and philosophy of Zen Buddhism. The dream, however, seemed to signify a return to her family religion, though the presence of the cross-legged meditative posture does not suggest a rejection of Zen. This posture in a monastic cloister creates an image of that which is contemplative in both the Christian and Buddhist traditions. The flowers seem to celebrate this reconciliation. Flowers are also indicative of the closeness of the archetypal Self and the unity that it represents. Johnson says:

> Such a symbol points to that archetype—the self—that transcends the opposites by revealing the central reality behind them and thereby unites them.
>
> Flowers are not only symbols of the feminine but also of the unified self: in Christianity, the rose that represents Christ; in Eastern religions, the thousand-petaled lotus that portrays the One.[163]

Johnson believes that the gift of a notable dream, such as this one, deserves some type of ritual. Not knowing what else to do, and with this in mind, the woman picked some flowers similar to those in her dream, took them to the ocean, and cast them onto the waves, "giving the gift she had received back to Mother Earth, back to the feminine sea of the unconscious."

Upon returning home, the woman was surprised to find a friend waiting, one who did not frequently visit. On a drive together, they passed a monastery. The woman was startled because she felt as though she had been in a monastery ever since her dream of the previous night. To her surprise, her friend was one of the few lay-persons given access to the cloister. She even had a key to the gate and suggested that they stop for a visit. Upon entering the chapel, the woman felt she had walked back into her dream! Every detail was just as it had been the night before. Placing herself in a meditative posture, she was able to recreate the entire mood of the dream, which swept back over her, filling her with serenity. In a short time she was given permission to visit on her own, to meditate and spend quiet time there nourishing that part of her that gave impetus to the dream.

All of us experience dramatic episodes of synchronicity from time to time, but many synchronistic coincidences are not spectacular. The hand of the Trickster may move with subtlety in our day-to-day lives, begetting synchronistic coincidences that take their place beside spontaneous thoughts, fantasies, and dream images to form entire leitmotivs of related events. While the varied facets of our inner lives on the surface appear scattered and unconnected, on closer inspection we may find that they participate in a number of *miniprocesses*, to use Ira Progoff's term, that appear and reappear over time. [164] Mini-processes may include memories that come to mind unbidden, things previously seen but now noted with interest for the first time—desires to read certain books or see certain people—all focused on a theme that for some time remains unobserved. The use of a personal diary or journal, as Progoff suggests, may aid our recognition of minipro-cesses as they unfold. Whether or not you use a journal, however, regular attention to your inner life often allows these processes to become evident.

A miniprocess may run a natural course and dissipate itself, or it may become part of a larger braid formed of a number of interwoven miniprocesses comprising a major theme in one's life. Here, we are approaching what Progoff calls major life *units*, periods when one is focused on a particular activity such as pursuing a career, becoming an effective parent, or developing a spiritual quest. Each of us can map out our own lives in terms of such units. They evolve out of what Progoff calls our *core creativity*, the central nucleus of creativity in one's being. This latter concept is almost identical to Jung's notion of the archetypal Self, as it is the core creativity that gives overall direction to one's life, bringing all major activities into alignment.[165] The similarity is not surprising, considering that Progoff studied personally with Jung.

Unbidden additions to one's inner life that generate from the core creativity are termed *emergents*, a word taken from the late-nineteenth-century French philosopher Henri Bergson, who used it to refer to new and creative elements thrown up in the process of evolution. Progoff observes that emergents are "in addition to causality." They cannot be anticipated or planned. Though he makes no point of it, this phrasing from an author who wrote an important book on synchronicity—*Jung, Synchronicity, and Human Destiny*—suggests that emergents include synchronistic coincidences. Such a notion is consistent with the fact that emergents are rooted in unconscious creative processes. Our own observations suggest that emergents include many small coincidences, as well as occasional dramatic ones, which aid the flow of one's life. For instance, you may come across books, friends, poems, and other emergents that give support to your personal direction.

To pursue your core creativity is to make life choices on the basis of what feels most deeply rewarding and satisfying rather than what society dictates or what others think is best for us. This is individuation. Joseph Campbell simply calls it "following your bliss."[166] In

The Power of Myth, a well-known series of interviews he made with journalist Bill Moyers shortly before his death, Campbell emphasized the importance of discovering exactly what it is that leads each of us to personal bliss or satisfaction. This can be anything: professional work of a particular sort, writing, painting, service to others, hiking alone in the mountains. The vital point is that we get it done; that we do not live our lives according to the directions of others, but follow the inspiration and promptings that arise from within and have deeply personal value.

This means that one must be able to say no to the incessant demands of an impersonal, bureaucratic world. Campbell points out that we are beset by a constant barrage of demands to become involved in more and more activities that give us no fundamental satisfaction. While such activities may seem to carry us further along in a profession or elevate us in the community, they may ultimately leave us empty, betraying us to a desolate inner life. They may mean nothing to the unconscious and instead carry us further and further from our true being, from our core creativity, from the archetypal Self. If we are to follow the path of individuation, we must make conscious choices about what is truly of value and what is not.

Both Campbell and Moyers point out that when you choose to follow your bliss, when you make choices based upon an inner sense of fulfillment rather than outer demands, there is often a sense of "hidden hands," of unexpected opportunities and unanticipated resources. This is synchronicity in the service of individuation. It is the influence of the Self in the world of human affairs that makes itself felt when we submit to the deep call of our being. Carl Jung termed this the *law of synchronicity*, meaning that when we are in accord with an archetypal process, then that archetype, in this case the Self, can influence events around us even at a distance, as in the old Taoist saying, "The right man sitting in his house and thinking the right thought will be heard a hundred miles distant."

Even without a conscious choice to follow your bliss, at critical moments in our lives the unconscious sometimes takes things into its own hands, as it were, in the form of synchronistic coincidences that trigger new phases in our lives. In *The Shared Heart*, author Joyce Vissell tells of her decision to marry her husband Barry, a decision that has led to a remarkably creative relationship but one that occurred only after much consternation over their both coming from different religious backgrounds. At one point, this conflict even led to their decision not to see each other again. At the point when things seemed hopeless, the mother of one of Joyce's friends felt an impulse to give her a poem from a prayer book. Underlined in the poem was the passage, "Above all else Love is important." This was evidently enough to trigger a major shift in Joyce's perspective, which she was able to communicate to Barry. They were married shortly afterward. [167]

Sometimes the obvious influence of coincidence on our lives is even more dramatic. Progoff notes that Abraham Lincoln's career in law, and so ultimately his presidency, is due in part to the unexpected acquisition of a barrelful of odds and ends, including a complete edition of Blackstone's *Commentaries*—a set of books that played a major role in the development of his interest in law. [168] We saw at the beginning of this book the impact of fortuitous coincidence upon the lives of Winston Churchill and Adolf Hitler.

Indeed, there are many instances of lives actually being saved by a fortuitous happenstance that stretches the limits of meaningless chance. In *The Challenge of Chance*, Arthur Koestler recounts a story concerning the British actor Sir Alec Guinness, who overslept one Sunday morning (despite the fact that he used two alarm clocks) and missed his regular train from London to his home near Petersfield. That morning there was a railway accident which involved that train and, in fact, the very car in which he would have been riding. This was a doubly unusual situation in that when Guinness first awoke he misread the time, thinking that he had not overslept at all. Were it not for this, he might well have decided to skip his regular attendance of mass in order to catch his usual train. As it happened, he went to the 9 A.M. mass thinking all the while that it was the 8 A.M. service, and then proceeded to catch a later train.[169]

Frequent news items tell of people, especially infants, who survive severe car accidents or long falls virtually unscratched. Interestingly, a survey of twenty-eight train accidents that took place between 1950 and 1955 revealed that significantly fewer people rode the trains on the days of the accidents than on comparable days in earlier weeks and months.[170] These individuals, for whatever reasons, did not make it to the train on the days of the accidents. The March 1950 issue of *Life* magazine reported that the entire membership of a church choir in Beatrice, Nebraska, arrived late for practice one evening. Five minutes after the scheduled meeting time of 7:20 P.M., the building exploded. The minister and his wife were delayed while she ironed their daughter's dress. One member needed to complete a geometry problem, another had difficulty waking her daughter from a nap, another waited to finish hearing a radio program, one couldn't start her car, and so on. Warren Weaver, in *Lady Luck: The Theory of Probability*, estimated the probability of this incident at less than 1 in 1,000,000.[171]

Returning to the theme of individuation, the following series of incidents, reported by an acquaintance of the ours, reveal the influence of the archetypal Self, Progoff's core creativity, expressed in a variety of ways, including a dream, a synchronistic coincidence, and a spontaneously written poem. These separate elements ultimately form a single pattern in one person's life. The story begins some years ago when the acquaintance was working near a beach where he saw hang gliders daily. They triggered something deep in him, because he dreamed shortly afterward that he was standing on the edge of a cliff with wings like a bird. Rather than overlooking the sea, however, he faced a desert with a horizon lit by the glow of a predawn sky. It was as if he were on the verge of flying up over the sand into the dawn. Two days later, he received a shipment of books from his mother. Opening it, he found a plaque on top with the likeness of a winged man standing on a ridge as if to take flight. The theme of flying, and especially flying toward the sunrise, was not new to him. Two years before, shortly after returning from a vision quest—a kind of spiritual retreat in the desert—he was moved to write a poem which, it turns out, contains much of the same symbolism:

> As the body becomes motionless,
> The urge to fly moves deep within.
> The Sky-Bird flutters.
> Towards the East, the sky grows brighter.
> Flying over the tattered masses,
> The Medicine Bearer searches out
>   the Living Waters.

He remarked that as outer activities become silent there begins the motion of an inner sense of soaring. The poem reminded him of the "Hermetic ascent" which, according to the esoteric tradition attributed to Hermes Trismegistus of second-century Alexandria, is the upward path of the soul when released from the confinement of the body. It is a clear instance of Hermetic journeying as described in Chapter 5. The tattered masses of the poem are the desert of the dream, and in both instances the ascent is made into the light of the dawn sky, signifying the coming of the sun.

Behind this series of events lies a spiritual impetus. The sun, as we have seen, symbolizes the archetype of the Self, but it also symbolizes the highest pinnacle of the human spirit, the *atman*. For Jung, the Self is the origin of the whole personality: "it is the relation or identity of the person with the superpersonal atman."[172] The Self provides a perspective, a line of vision in the direction of the greater transpersonal Self, the atman. The water in the poem points to both the collective unconscious, the wellspring of all archetypes, and to the nourishment of the ascendant spirit.

Spiritual Work

In the preceding account we find the guide of souls playing his role in the context of a spiritual aspiration. Many similar accounts suggest that the archetypal guide's synchronistic gifts can be important in spiritual work. If one is authentically called to such work, then it involves individuation in the sense that one is guided toward the unity that lies beyond all opposites by applying oneself on a particular spiritual path. The act of discipline arising naturally from the Self resolves the conflict of the opposites. As in other instances of individuation, the play of the Trickster may assist the individual's quest in a variety of ways. For instance, in many traditions there is a belief that when the time is right the student will be found by a teacher, a seemingly synchronistic meeting that we may consider an instance of Jung's law of synchronicity.

And consider the advice of the old alchemist to one of his disciples: "No matter how isolated you are and how lonely you feel, if you do your work truly and conscientiously, unknown friends will come and seek you."[173] In this connection, on an inner level, alchemy itself seems to have been an ancient form of individuation. Since its inception in ancient Egypt, (if not earlier) alchemy has been linked with the transformation of matter into a higher or divine nature. Its historical roots go back at least to the process of mummification by which the physical substance of the deceased's body was thought to be transformed into a sublime but still physical substance, essentially by a magical process.[174]

In the Middle Ages, the goal of alchemy was to transform lead or some other ordinary substance via a sublime material known as the philosophers' stone into gold. The medieval alchemist, however, was quite different from his modern scientific counterpart. Rather than valuing independence and separation from the object of investigation so important to modern mechanistic science, the alchemist was subjectively involved in an essential way. Historian Morris Berman points out clearly in *The Reenchantment of the World* that success in the alchemical enterprise involved an intense personal and contemplative, rather than impersonal and objective, involvement with one's work.[175] Indeed, for the medieval mind, the notion of separating the outer work, carried out by the various utensils, instruments, and chemicals in the laboratory, from one's own inner nature, was a foreign idea. In psychological terms, the outer work of transforming chemical elements was paralleled by an inner work of transforming oneself; to alchemists, it was the same work. Jung studied intensively the texts left by medieval alchemists, and it is apparent in his scholarship that many alchemical writings yield a double meaning, the outer one referring to chemical processes and material substances and the inner one dealing with the transformation of the alchemist's Self. In this latter transformation, the philosophers' stone may be thought of as symbolizing the archetypal Self.

The transmutation of the base material of the alchemist's own psyche via the sublime substance of the philosophers' stone represented the spiritual completion or individuation of his being. The stone itself was said to have the miraculous power to transform other material via its own nature. Thus, if one could create even the smallest amount of the stone, it could be used to transform lead, for example, into gold. Such an idea is more than a little fantastic if taken in literal terms, but as a metaphor of the inner transformation of the alchemist it holds a magnificent lesson: once the great work is accomplished and a person truly lives from the inner Self rather than from the ego, his presence radiates a powerful, transforming effect on all around him. This wisdom is well known in the East, where an important part of spiritual work is simply to *be* with the teacher as often and as much as possible. The power of his presence alone is healing.

Von Franz notes a beautiful instance of this wisdom in the series of ox herd drawings of Zen tradition. In them, the ox herd sets out to find the lost ox, which represents his Buddha nature, and, symbolically, the Self. After a considerable time he finally finds it and rides it home. In the last picture of the series the ox herd is often shown as a wise old man with a sort of banal, friendly smile and a begging bowl. The accompanying poem reads, "He has forgotten the gods, he has even forgotten his enlightenment. Quite simply he goes to the market place begging but wherever he goes the cherry trees blossom." Here, von Franz comments, "we see the healing effect on outer things, even on nature."[176]

In the spiritual quest, as in other aspects of life, the gifts of the Trickster may not always seem at first to be helpful. For example, in suggesting that one choose a special goal each day to work toward, Swami Rama humorously observed that "The day you say, 'I will love everyone and not hate anyone today,' you will find that all your enemies are coming to you. They come via telephone calls, or in letters, or you hear someone talking about you."[177] This, of course, is exactly what is needed for growth, and it is the kind of thing the Trickster is cut out for.

Synchronistic gifts, however, are often more congenial, especially as progress is made along the path. The Mother, who was a great yogi and a colleague of Sri Aurobindo, once remarked:

If you have within you . . . [an inner Being] . . . sufficiently awake to watch over you, to prepare your path, it can draw towards you things which help you, draw people, books, circumstances, all sorts of little coincidences which come to you as though brought by some benevolent will and give you an indication, a help, a support to take decisions and turn you in the right direction. But once you have taken this decision, once you have decided to find the truth of your being, once you start sincerely on the road, then everything seems to conspire to help you in your advance.[178]

A careful reading of the works of The Mother and Sri Aurobindo strongly suggests that the inner Being referred to above is, not surprisingly, the archetypal Self as its influence ascends during individuation; that is, as one submits to one's destiny.

Those who have had the opportunity of spending time with accomplished sages have observed that they are surrounded by frequent and helpful coincidences. When asked about this, one sage replied, "It is the cooperation of nature," a comment poetically reminiscent of the epitaph to the final ox herd drawing. The answer implies an absence of the usual attitude of attempting to control the events of the objective world. Rather, it points to an attitude of synergy by which a state of cooperation exists between the individual and the world. Ram Dass, the well-known spokesman of Eastern spirituality, once commented to one of the authors that the frequent "miracles" that surrounded his own teacher, Neem Karoli Baba, were not so much a matter of outward control but rather the result of an effortless harmony between himself and the world. He did not identify himself with the process within the skin, as most of us do, but with a larger reality. The discipline of meditation tends to soften boundaries and, as we have noted, act as a catalyst to synchronicity.

The basic notion that at higher states of human development one may enter into a kind of synergistic unity with the world was expressed by no one more eloquently than Swami Rama Tirtha. In a talk given at the Golden Gate Hall in San Francisco in 1906, he said:

> The truth is undeniable that so long as you are in perfect harmony with nature, so long as your mind is in tune with the universe and you are feeling and realizing your oneness with each and all, all the circumstances and surroundings, even winds and waves, will be in your favor. . . . Bear in mind that the Self in you is the same as the Self in all surroundings or environments and when your mind is in harmonious vibration with this underlying Self Supreme and your body has become the whole world, outside aids and helps must fly to you.[179]

Rama Tirtha first set foot in the United States in 1902. As he was disembarking from a steamer in San Francisco, a curious American noticed that he was not in the usual rush to get ashore and asked, "Where is your luggage, Sir?"

"I carry no luggage," the Swami replied, "but what I have about me."

"Where do you keep your money?"

"I keep no money."

"How do you live?"

"I only live by loving all. When I am thirsty there is always one with a cup of water for me, and when I am hungry there is always one with a loaf of bread."

"But have you, then, any friends in America?"

"Ah, yes, I know one American—you. . . ."[180] They soon became companions.

THE TRICKSTER AT PLAY

For most of us, unfortunately, life is more difficult. How are we best to understand synchronicity in our day-to-day lives? And how can we honor its master, the Trickster, in our thought, feeling, and action? A starting point is to realize that above all else the Trickster is playful. In his book *The Return of the Goddess*, Jungian analyst Edward Whitmont notes that the key to appreciating play is having an attitude of openness. He writes that

> [Play] is at its best when performed for its own sake, not for any purpose or achievement other than itself. Play is self-discovery in the here and now. It is spontaneous, yet has its own discipline. It is light, yet potentially passionate. It is discovery, and it is enjoyment of one's own and of others' possibilities, capacities, and limitations. Most, if not all, great discoveries, even in science, have been the result of intense effort along with playful curiosity and joy of exploration on the part of the discoverer.[181]

Whitmont observes further that "play is the Yin side of exploration, just as exploration, experimentation, and discovery comprise the Yang side of play, enjoyment, and feeling."[182] That is, play may be thought of as the feminine side of the masculine enterprise of exploration and discovery, while these enterprises are the masculine side of the feminine activity of play.

For the inner, archetypal Trickster, play includes a synchronistic taking hold of whatever materials come to hand in order to break the boundaries of our usual perceptions of reality. In addition, trickster stories almost universally emphasize his doing exactly what he pleases regardless of the consequences. The apparent selfishness is, in part, a way of portraying his sovereign nature as an uncontrollable aspect of the human psyche that originates totally outside the reach of the conscious mind. The meaning of his actions, however, depends not on himself but on some deeper aspect of the psyche in whose service he acts. Whether the Trickster confronts us with the contents of our shadow or tosses up synchronicities that speed us toward the fulfillment of our destiny, in a final sense he acts under the direction of the archetypal Self.

Obviously the Trickster is not a moral hero. We must keep this fact in mind in coming to understand his synchronistic activities. Many stories attest to this absence of morality; Barry Lopez's highly readable book, *Giving Birth to Thunder, Sleeping with his Daughter*, about the American Indian Trickster Coyote, contains several. These and many other trickster stories are scatological, often emphasizing the Trickster's greediness for sex or his breaking of all sorts of taboos. The Trickster's absence of morality means that his span of activity is not limited by any notion of fair play. Even when we understand his service to the archetypal Self, we may sometimes find his synchronistic gifts much to our dislike. John Lilly once observed, in speaking of synchronicity, "Cosmic Love is absolutely ruthless and highly indifferent: it teaches its lessons whether you like/dislike them or not."[183]

On a psychological level, the amorality of the Trickster represents his utter disregard for one's state of mind at the time that a synchronistic episode is enacted. If we are angry he will do something to infuriate us! If we are happy, he may still do something to infuriate us, or he may live up to this title of delight-maker and do something wonderful. We may be firmly convinced that we are absolutely right about some point that we are boldly presenting to others, and he will do something that shows us to be fools! There are no limits to his antics. It is his delight to shatter our boundaries, borders, and frames, stripping us of our protective coloration and baring us helplessly to something new. This is his play, and when we ourselves are playful, we are in harmony with him.

The Mood of Play

Adopting a playful mood toward synchronicity means following the Trickster wherever he leads, knowing that we are led by the guide of souls. It means to lighten up—to pay attention to where the flow of coincidence leads. In doing so we honor the Trickster. This does not mean that we should throw ourselves squarely in the path of coincidence, accepting every chance event that befalls us as some divine gesture. To do so would be foolish indeed, for our benefactor is, after all, a prankster, and enjoys nothing better than making us play the fool. Being observant and alert is more important than blind submission. If chance plays the prankster, perhaps we are in a hurried and anxious mood, packing our urgent frustrations into the shadow where the Trickster is gaming with them. Perhaps synchronicity is showing us some new facet of our development by leading us to a certain book, an unexpected friend, or the possibility of a new career. In such instances, one suspects the involvement of the archetypal Self and would be wise to remain alert to further cues, not only in the form of synchronistic coincidences but also in dreams and fantasies.

It is important in all this to realize that synchronicity in its largest sense is not restricted to external coincidences but includes the inner subjective gifts of the imagination: fantasies, unannounced inspirations, and feelings. As we have seen, Hermes is himself lord of the imagination as it comes like a gift from beyond the borders of the conscious mind. Allowing our imagination to play, letting our fantasies have their day, is to honor him. Utterly to deny this natural tendency of the mind, to suppress the imagination, to refuse to give it a hearing, to refuse even to honor it with our momentary attention, will cause it to carry its case to the shadow where the sympathetic ear of the prankster awaits it.

Allowing the imagination to play means to lighten up from time to time, to let our fantasies run free. To do this we must relax rigid attitudes or moods, even perhaps our concepts of morality. This does not mean acting out immoral wishes but consciously following the imagination in order to help free ourselves from being the victim of the Trickster's pranks. When we allow this play, the Trickster brings insights about our unconscious hopes, fears, and passions. In doing so he frees us from the compulsion to act out motives we do not understand. The acting out of unconscious motives is the opposite of true play; rather, it means possession by an archetype and therefore the overwhelming of the ego, a dangerous form of psychological blindness.

Even play makes its demands. It asks that we temporarily admit *any* possibility even if it is immoral. Temporarily relaxing your morality means putting aside your culturally created and therefore limited conception of reality, including the reality of your own self. The Trickster can then reveal aspects of our selves that are hidden from our scrutiny. Growth of the personality is certainly not guaranteed by this. But if we allow the Trickster to be our guide and we follow his play consciously, we are given the very real possibility of expanding our sense of who and what we are.

Allowing true freedom to the imagination requires that we take the courage to bare ourselves to an insecurity that comes with giving oneself over to the irrational. Flights of imagination may threaten deeply rooted attitudes or moods. Such attitudes are comfortable, and they impart a sense of familiarity to everyday living. We feel their rightness, and so we may find it disquieting to open ourselves wholeheartedly to the irrational play of the imagination. Fortunately, we are to some degree protected by the spirit and atmosphere of play itself, which are intense but light and good-humored. These protect the players from taking themselves too seriously. They frame play activity, setting it aside from other, more serious, endeavors. Thus play, the activity of the unbridled Trickster, is, in this paradoxical sense, bounded.

If we wish, we may strengthen the separation of play from the serious business of daily living by selecting a special place for it. An artist or writer may design and decorate a particular room in order to stimulate the imagination and further her creativity. Or you might choose particular clothes that make you feel especially relaxed, comfortable, and at ease with your imagination. Or you might go for a walk to let your thoughts flow freely. These strategies lower our anxieties about letting go and allow our imaginations free play. They also mark a distinct beginning and end to the period of imaginative play, so that we may easily return afterward to more mundane activities.

Framing play as something set aside from more serious endeavors gives it a ritual quality. Rituals typically have a formality that separates them from the affairs of everyday life. As Whitmont notes, however, rituals also serve a very special role by ordering and binding together whatever they concern.[184] For example, the marriage ceremony binds together two people in the ordered relationship of marriage. The word *ritual*, in fact, comes from an Indo-European root which means "to fit together." It is related to such words as art, skill, order, weaving, and arithmetic, all of which involve fitting things together to create order.

What characterizes the play of the Trickster is its unexpectedness, the surprising combinations of reality that can pop up in his hands. Yet from our point of view rather than his, play is a conscious following of the disorder created by him, so that we ourselves may create order. We do so by trying to understand the implications of his activities, much as we might try to understand the implications of our dreams: Do they tell us something about our shadow? Do they seem directed by the higher agency of the Self? What might they mean in terms of our individual growth?

The two ritual aspects of play, the fact that it is set aside from the serious affairs of living and that it serves the role of creating order, are both seen clearly in a particular method of self-exploration termed *active imagination* and developed by Jung himself, who used it extensively. In it, fantasy images are allowed to take on a life of their own, then to interact with us as if they were real. Fantasy figures may share their concerns and interests, sometimes leading us to their own worlds to show us things they wish us to see. Robert Johnson, in his book *Inner Work*, gives a clear account of this wonderful procedure for expanding self-knowledge by direct experience of aspects of ourselves that usually remain hidden. People working with active imagination often take full advantage of the ritual aspects of play by setting aside a special time or place for it. In this way they find it easier to leave the sometimes alluring fascination of the fantasy images.

Life as Play

It is possible, if we have real courage, to live all of life as if in play. This does not mean being frivolous or lacking compassion toward others. It means to carry a light, trusting, and open attitude toward ourselves and the world. In a fourteenth-century B.C. Egyptian papyrus depicting the final judgment of a deceased person, his heart is seen to rest on a tray at the end of one arm of a great balance. On the tray at the opposite end of the balance rests a peacock feather. Below stands Amemet, devourer of souls, ready to take his soul if he fails to pass this judgment. On the right is Osiris, lord of the underworld, ready to take it if he passes. The message here seems all too clear. Lighten up! To pass the judgment one's heart must be lighter than a feather.

In Tibetan Buddhism it is said that what distinguishes human nature from that of animals is not intelligence but humor. To experience life as play one must learn to see with the eyes of humor. This helps us balance the tragedy of human existence with the wonder of it. Such an attitude requires courage because it demands that we open ourselves both to uncertainty in the outer world and to the irrational in the inner world.

A truly playful attitude, even if short-lived, can act as a catalyst to synchronicity. Moreover, an attitude of lighthearted openness reduces the shadow to a bare minimum, since the defenses are relaxed. As a consequence, coincidences are often delightful. At times, a positive sense of trust and openness will allow everyday problems almost to solve themselves, as opposed to the more usual sense of struggle against chance events that the Trickster so often throws in our path. Scientist and visionary John Lilly once observed:

> There are days in which all events planned for the day, for the next week, for the next few months, line up, almost automatically resolving former conflicts of hours of meetings, deadlines, financing. Someone (A) telephones, asking for a meeting a week hence: one writes the date into the engagement book. A crisis develops negating that date. Within a few hours person A calls and asks for a change of date because some factor changed in his/her life apparently unrelated to one's own. Thus one is given the time to resolve the crisis.[185]

Lilly goes on to suggest that one should expect the unexpected every minute of every hour of every day.

Opening the mind to a lighthearted, playful attitude, we may avail ourselves of intuition, which is a particular kind of gnosis, or knowledge, that seems to come through the now permeable borders of the conscious mind. Intuition is a type of knowledge emphasized in virtually all spiritual traditions. This is not to say that to be lighthearted is to become psychic, as the term is usually used, but rather that we may develop an exquisite feeling for certain situations, a feeling which, if trusted, often proves correct. Intuitive feelings hold a special relationship to synchronicity, a relationship that a few people have actively cultivated.

One who has is Peter Caddy, one of the three original founders of the Findhorn alternative community in Scotland, who developed his sense of intuition during many years of training. He recounts a particular sequence of events, all of which happened because he followed his own intuitive promptings without hesitation. Halfway through a cup of coffee in a café in Oban, Scotland, with only a shilling in his pocket, Caddy felt a prompting to go help a friend who was hundreds of miles away in London. Walking immediately out the door he spotted a Rover approaching and asked the lady driving for a ride. She agreed. She was going all the way to London, carrying a hamper with an entire cooked chicken which they both enjoyed. Later, after completing his short stay, Caddy caught a bus out of London. At a traffic light he spotted a sports car with a vacant seat. Acting again on a prompting, he got off the bus and asked the driver of the sports car if she was going north. She was, indeed, and she took him all the way to Scotland at eighty to ninety miles per hour! She had lots of sandwiches with her, so again there was plenty of food. Late that night, and with some distance still to go, he caught a ride with a truck transporting fish. The driver had been on the road for sixteen hours and was glad when Caddy offered to drive. In return, the driver bought breakfast for both of them. Later, Caddy commented, "The whole journey had been accomplished in less than four days. I could not have done it as quickly in my own car, and in spite of starting off with only a shilling everything had been provided through instant obedience and having the discipline to follow through."[186]

In Peter Caddy's story we see the remarkable possibilities of the cultivation of intuition combined with a disciplined submission to its promptings. We also see a person who is willing to let go of his own seemingly urgent needs and projects to immediately follow and fulfill the most subtle nudge of an inner voice. He is available, in a sense, for his destiny to be molded effortlessly by the outer situation and his exquisite inner response to it.

A Zen story illustrates what it means to live with one's whole being open to whatever destiny brings to it. It concerns a Zen master, Hakuin by name, and a young unmarried woman who attended his temple. She became pregnant, and when she was pressed by her angry father to reveal the name of her lover, she named Hakuin as the father. The girl's father stormed into the temple carrying the infant, placed it at Hakuin's feet and rebuked him for his scandalous conduct. Hakuin replied, "Oh, is that so?," picking up the child and placing it in his tattered robe. Soon word got around that the revered master was caring for his own illegitimate child, and indeed he could be seen carrying the baby in his arms when he made the rounds of the town with his begging bowl. Naturally, such conduct on the part of the master was a scandal, and his disciples began to leave him. The town filled with gossip about him. Of all this, however, Hakuin took no notice. The episode was finally resolved when the girl, in anguish for the loss of her child, admitted to her father that Hakuin had not been her lover. In mortal terror for his soul, the girl's father rushed to the temple to beg for Hakuin's forgiveness and the return of the child. Hakuin's reply was simply, "Oh, is that so?," and he handed over the infant.[187]

This story exemplifies the master's connection through the Trickster to the archetypal Self and the openness and confidence—faith would be a better word—that this connection grants, an attitude we also saw in Peter Caddy. It also demonstrates a flexibility that embraces whatever life offers. Thus, in a metaphorical sense Hakuin became able, like the Trickster himself, to take any shape; he could either become or cease to be father to the unmarried girl's child.

This ability—to take any shape effortlessly—was also a goal of medieval alchemy. It was thought to be achieved through a psychological death and resurrection symbolized by the resurrected god of the ancient Egyptian underworld, Osiris:

> The highest goal of the resurrection was thought of as this ability to be completely free to change into any shape . . . the alchemists connected this idea with their concept of the philosopher's stone, that divine nucleus in man which is immortal and ubiquitous and able to penetrate any material object. It is an experience of something immortal lasting beyond physical death.[188]

As we have seen, the philosophers' stone symbolizes the archetypal Self. Inwardly, when its influence on the personality is great, it tends to make one immune to the harsh inflections of reality, so that—in Jung's own words—"a personality develops that suffers only in the lower stories, so to speak, but in the upper stories is singularly detached from painful as well as joyful events."[189] Outwardly, the influence of the Self may be felt synchronistically in one's entire sphere of activity. Von Franz comments:

> Because one is in connection with the self, the self begins to have a certain effect. . . . *If one is connected with the self inwardly, then one can penetrate all life situations.* Inasmuch as one is not caught in

them, one walks through them; that means there is an innermost nucleus of the personality which remains detached, so that even if the most horrible things happen to one, the first reaction is not a thought, or a physical reaction, but rather an interest in the meaning.[190]

True openness to experience comes via a connection through the Trickster to the archetypal Self. This openness is play, and play is the Trickster's game—irrational, paradoxical, and creative.

THE DIVINE TRICKSTER

"Gaiety, love, and sweet sleep" are the ravishments which flow from Hermes' playing of the lyre.[191]

On certain ancient vase paintings Hermes was represented as identical with the silenoi, creatures that resembled satyrs and worshiped Dionysus, the god of wine, divine intoxication, and dance. They were themselves heavily inebriated most of the time! This association with the half-human silenoi, and their involvement in turn with Dionysus, suggests two major facets of Hermes and his gift of synchronicity.

First, it connects him with the psychoid level of the unconscious where the animal roots of the soul have their origins. Through this connection the divine Trickster weaves a bridge between the highest Olympian spheres and the unknowable depths of the unconscious, "conjuring . . . luminous life out of the dark abyss that each in his own way is."[192] As we have seen, Carl Jung believed that consciousness cannot exist by its own energies alone but relies on the unconscious for the breath of life. In enriching our connection with the unconscious through imagination and synchronicity, Hermes adds to our lot of this vitalizing breath. His association with the psychoid level establishes synchronicity as a link between the world of the mind and the world of matter. This level is the very ground of being, the *unus mundus*, the meeting place of mind and nature. It is this level, if Bohm is correct, that may give birth to archetypal patterns of cosmic extent, experienced both in the depths of the mind and in the larger universe. From this deep well synchronicity draws its meaning, so that each synchronistic coincidence mirrors the same significance in the world of objective matter as in the world of inner experience.

Further, the association of Hermes with the silenoi, and through them with Dionysus, allies him with divine intoxication, with music and with dance. Thus we discover that Hermes, inventor of the lyre and the world's first musician, stands along with Dionysus and the Indian god Shiva at the center of one of the most ancient and elevated concepts of creation, a concept based in rhythm, movement, and mutual attraction; in a word, the dance. The Greek writer Lucian wrote, "It seems that the dance appeared at the beginning of all things . . . since we can see this first dance clearly appear in the ballet of the constellations and in the lower lapping movements of the planets and stars and their relationships, in an ordered harmony."[193] In Indian mythology the divine Shiva is known as the lord of the dance, and all creation is his theater. It is his dance that resonates throughout the universe as the rhythmic energy that is at the foundation of everything.

The mood of the dance, like that of play, is one of dynamic tension between discipline on the one side and self-abandonment on the other. Such a mood aptly describes the ambience of our relationship to the world when we are most open to experiencing synchronicity; we play, we dance. Let us stop briefly to again examine the mystery of synchronicity; then we will return to this idea.

Beyond all our best efforts to understand it, in this book and elsewhere, synchronicity still embraces a deep enigma. Like certain processes in quantum physics, synchronistic coincidences speak to us about existence in the language of impenetrable conundrums. They are reminiscent of Zen koans, riddles for which there is no rational solution, even in principle. What is the sound of one hand clapping? What was the appearance of your face before the universe began? Such riddles place the Zen student in the position of the proverbial flea attempting to bite the iron bull. It cannot be done! The flea must eventually give up in abject frustration in the face of unyielding and enigmatic reality. Like the flea locked in its impossible task, we also face an impenetrable wall when we confront synchronicity. As when trying to explain the probability wave function of the quantum potential, we can draw sketches of explanations in the air, but ultimately we are slowly drawn deeper into its riddle.

The inescapable insinuation of synchronicity, however, is that the cosmos is undergirded by teleology. Synchronicity reminds us of this order and beckons us to enter into it. Purpose in the form of synchronistic coincidences finds us even in the banalities of our daily routines. This is not the Logos, the idea of a universal order that endows the cosmos with the semblance of a rational mind, that was subscribed to widely in the late classical world of Greece and Rome. It is rather the order of a trickster, fraught with the unpredictable— a joker's brew of the unexpected and unforeseeable. Its purpose cannot in the end be grasped with the rational mind. It must be lived with one's whole being.

The lesson of synchronicity is that it must be grasped with an open hand. The examples of Rama Tirtha, Peter Caddy, and the Zen master Hakuin all illustrate faith in this openness. The mind seeks knowledge as a handle on the cosmos from which it gains leverage for greater control and self-satisfaction, but synchronicity will not allow itself to be used so roughly. Like the flea, we must eventually give up our effort to penetrate the impenetrable and surrender to a reality which we cannot master but to which we must submit. In the end, to be honest to our exploration of synchronicity we must ourselves surrender to it. This means to relax and allow the sometimes fickle tide of fate to take its natural course, to let it wash over and benignly carry us. We must sacrifice the urgent, petty agendas of the ego to a larger field of participation. We must learn humility and own humor, finding guidance in intuition and making logic a servant rather than a master. Control is a personal experience, surrender is a transpersonal one. Through surrender we learn to move with the rhythms that flow through our existence and in so doing open ourselves to the wellsprings of life that are the gift of the divine Trickster.

Dance, like play, is a metaphor for a state of being that is both relaxed and disciplined. Both are open and responsive to relaxed intuition and sensitive to the ambience of our entire situation, its rhythms, its melodies, its tragedies, its humor. To dance is to move in the rhythm of this entire orchestration. And so we must learn to dance.

> *And if you would understand what I am, know this: all that I have said I have uttered playfully, and I was by no means ashamed thereby. I danced.*
>
> JESUS CHRIST
> *Acts of John*[194]

APPENDIX I
OMENS AND DIVINATION

> *. . . the Thriae [mountain goddesses] showed*
> *Hermes how to foretell the future from the*
> *dance of pebbles in a basin of water; and he*
> *himself invented both the game of knuckle-*
> *bones and the art of divining by them.*
>
> ROBERT GRAVES
> *The Greek Myths*

Omens are events—usually natural events such as a flight of birds or a clap of thunder—which act as signs of things to come. Divination requires active participation. One must pose a question, then toss yarrow stalks or coins, fire a tortoise-shell to make a pattern of cracks, or in some other way create a display of seemingly random events in which an answer can be read. We will begin with omens.

We most often think of synchronistic coincidences as precipitated by previously active, unconscious archetypes. Thus, the appearance of the scarab beetle was preceded by the earlier emergence of the symbolic beetle in the mind of Jung's patient—leading to her dream of this insect. Likewise, a series of appearances of fish seen by Jung over several consecutive days was preceded by his intense involvement in the study of the meaning of the fish symbol. If the situation were reversed, and the outer event were seen first, then we would have an omen.

As we noted in Chapter 5, the second example described by Jung in his essay on synchronicity involved an omen. This was an incident in which one of Jung's patients was struck down with a fatal heart attack at the time that his wife, at home, had become distressed because of a large gathering of birds outside the house, an event that had also heralded the deaths of her mother and grandmother. We also noted the ancient association of birds with the soul and the flight of the soul at death as well as the observation that omens frequently involve natural events, as opposed to most synchronistic coincidences which seem to consist of human creations such as numbers or words.

OMENS AND SHAMANISM

It is possible that omens, with their intimate connection to the world of nature, were among the first meaningful coincidences experienced by early humankind. Many anthropologists and psychologists have considered in detail the way early humans experienced their world.[195] They speculate that, like young children, people lived in a world of magic. The persons of knowledge in this world were magicians. Images of these practitioners of magic can still be found on the walls of caves that were used by Paleolithic tribes. Joseph Campbell notes, for example, that no less than fifty-five images of magicians may be located among the herds of grazing animals on the walls of the various caves. These ancient wizards were the early counterparts of modern shamans.

Shamans, both ancient and modern, provide access to the world of spirits for their communities, and they also function as physicians. They make use of a deeper understanding and mastery of life and its mysteries than do those who have not been chosen for their special training. It is likely that in ancient cultures it was the job of the shamans to interpret the meanings of natural signs or omens—to see the future in flights of birds, patterns of waves, and shifting shapes of clouds.

As we have seen, however, omens and synchronicity are two sides of the same coin. Thus it would not be surprising to find synchronistic coincidences in association with shamanism. Looked at from the other side, the unique training and skills of the shaman might well be expected to engender a state of mind, and brain, that catalyzes synchronicity. Michael Harner, in his delightful book, *The Way of the Shaman*, says:

> In the shamanic work it is important to be on the lookout for the occurrence of positive synchronicities, for they are the signals that powers are working to produce effects far beyond the normal bounds of probability. In fact, watch for the frequency of positive synchronicities as a kind of homing beacon analogous to a radio directional signal to indicate that right procedures and methods are being employed.[196]

DIVINATION—ANCIENT AND MODERN

In the great farming cultures of the ancient Middle East, two types of divination practices were widely performed. One involved a shift in consciousness to a trancelike state, produced by any of a variety of methods such as peering into a clear crystal ball or looking into the smooth surface of a lake as it reflects the sky. The ancient temple at Delphi apparently produced high concentrations of carbon dioxide which transformed the state of mind of the oracles who resided there. Under these conditions, one is subject to visions or other trance-induced phenomena. The other type, and the one of interest to us here, is the art of reading meaning in seemingly random events. The more random the event the better. This amounts to the reading of "signs," or omens.

In ancient Mesopotamia divination was almost an obsession. Virtually every event was thought to have some meaning. According to historian of the occult, Seligman:

> The Mesopotamians were masters in the arts of prescience, predicting the future from the livers and intestines of slaughtered animals; from fire and smoke, and from the brilliancy of precious stones; they foretold

events from the murmuring of springs and from the shape of plants. Trees spoke to them, as did serpents, "wisest of all animals." Monstrous births of animals and of men were believed to be portents, and dreams always found skillful interpreters. Atmospheric signs, rain, clouds, wind and lightning were interpreted as forebodings; the cracking of furniture and wooden panels foretold future events. . . . Flies and other insects, as well as dogs, were the carriers of occult messages.[197]

Mesopotamia was known throughout the ancient world for its great magi, or magicians. But divination was also practiced in other ancient civilizations and, indeed, would seem to have been the rule rather than the exception. The list is long and includes all of ancient Europe, Egypt, Japan, China, and Israel.

In ancient China it was common to engrave questions on smooth bone or tortoiseshell, then fire it and read the answers in the patterns of cracks caused by the heat.[198] A procedure practiced widely in the ancient world and carried right down through the Middle Ages was to sacrifice an animal—a pig, cow, goat, or often a bird—cut it open, and read the answer to questions in the patterns of the intestines.

While this sort of divination would, on the surface, seem to require a great deal of intuition, this was not always the case. Rules were often provided, and specific ones at that. Six Babylonian tablets, for example, give instructions for performing divination by means of pouring oil into a bowl of water, or alternatively water into oil:

> If the oil divides into two; for a campaign, the two camps will advance against each other; for treating a sick man, he will die. If from the . . . oil two drops come out, one big, the other small; the man's wife will bear a son; for a sick man, he will recover.[199]

Several such examples are found in Loewe and Blacker's collection, *Oracles and Divination*. For instance, in reading smoke, two tablets advise:

> If the smoke bunches toward the east and disappears towards the thighs of the baru [priest], you will prevail over your enemy. If the smoke moves to the right, not the left, you will prevail over your enemy. If

it moves to the left, not to the right, your enemy will prevail over you. [200]

One wonders if such canned procedures could possibly yield valid results. Perhaps if they did, the critical factor that made practitioners successful was their attitude in approaching the act of divination. The experience of the authors with the *I Ching*, an ancient Chinese form of divination widely used today, affirms that approaching the task with an attitude of respect and reverence is important in obtaining useful results.

We may never know whether these ancient procedures yielded valid results, but a certain insight into this question is provided by looking to a culture in which such practices have carried through into modern times. This is the culture of Tibet. Until the Chinese takeover, Tibetan culture was rich in practices of divination. Not only were there many types, ranging, for example, from seeing visions to the deciphering of bird behavior, but many people practiced the divining arts. Much divining was done by lamas (priests), but anyone could actually become a *mopa* or diviner. Lama Chime Radha, Rinpoche, comments, "The career of a professional diviner was a somewhat insecure way of supporting one's self and one's family. Anyone whose prophecies were not confirmed by events would quickly lose his reputation, and his trade would suffer accordingly." He goes on to say that "some diviners were trusted as being honest and genuine, while others had the reputation of being charlatans and were not respected."[201]

EVERYDAY DIVINATION

The reading of omens is still practiced by many people. Shamans are still to be found in traditional tribal cultures throughout the world. Divination, however, is not restricted only to those who possess special knowledge or powers. Fascinating instances of divination still abound among many cultural groups throughout the world, from Europe to

Appalachia. Methods include the reading of the meaning of patterns in the smoke of recently extinguished candles; deciphering messages in animal behavior (one of the most ancient of recorded divination practices); and those mentioned earlier.

Perhaps, however, one can make too much of hidden meanings and secret knowledge and thereby overlook events that seem to carry their own evident messages. For example, when an acquaintance of one of the authors was hoping to conceive a child, he and his wife frequently used home pregnancy tests—the type that can be purchased over the counter in a drugstore. When the results are positive some display a small ring about a centimeter in diameter floating in a little test tube. One morning, though the test disclosed nothing, the hopeful husband found himself with a particularly enthusiastic and numinous feeling. Driving to work that morning he noticed an unusual cloud. It formed nearly a perfect ring in an otherwise empty sky. It lasted for perhaps five minutes and seemed to transmit a joyous feeling that a child was coming. A few days later the test tube oracle followed suit and produced a fine ring of color.

In such cases omens clearly carry something in common with other instances of synchronicity: a feeling of power or numinosity, as if one were touched with divine authority. Such a feeling may transmit a sense of deep joy and trust, as it did in the above instance, or a sense of fear and foreboding, as in the instance of the birds that surrounded the house of the woman whose husband was at that moment suffering a fatal heart attack. The woman needed no special facility to interpret the omen, as she had experienced it twice before.

ACTIVE DIVINATION

Those interested in an active approach to divination can use various procedures. The ancients, for instance, released captive birds in order to observe their flight. Active methods of divination most widely practiced today are the *I Ching or Book of Changes* and the tarot.

Thumbing through the *I Ching*, we find a rich collection of proverbs and images taken both from the world of nature and from the political and social world of ancient China. Much of this material originated in the Confucian tradition, with its concern for harmonious political and social life. The roots of the book, however, stretch deep into the soil of Taoism, perhaps the most holistic of all worldviews. Taken as a whole, the book represents a high watermark in human wisdom and has been studied for its own sake.[202]

To consult the *I Ching*, you would keep in mind the question you wish to ask of it, while making six tosses of the fifty yarrow stalks (three coins are usually used in the West). The results of the tosses lead you to a particular set of parables in the book which, if you are successful, respond to the question you have asked.

A common first experience with the *I Ching* is one of amazement. The parables have an uncanny way of speaking directly to the heart of the question, even for the most skeptical. Moreover, the parable form of the answer encourages active exploration of its wisdom.

Jung was fascinated with the *I Ching* and frequently consulted it. His interest was amplified by his friendship with Richard Wilhelm, the prominent sinologist and the first to produce a thoroughly successful translation of the book into a Western language. It was in 1930 in his memorial address for Wilhelm, that Jung presented the idea of synchronicity as a concept "not based on a causal principle, but on a principle (hitherto unnamed because not met with by us) which I have tentatively called the synchronistic principle."[203] In that address he suggested that this principle underlies the action of the *I Ching*. Essentially, the idea was that the random tosses of the yarrow stalks or coins provide an opportunity for a pattern to emerge that reflects a much vaster pattern of events.

The West has its own counterpart to the *I Ching* in the tarot, a deck of seventy-eight cards representing various figures derived from medieval court life, though there is reason to believe that the origins of the deck may be much more ancient. With a question firmly in mind, the deck is shuffled and cut a number of times in a highly ordered fashion, and then the cards are laid out to be read.

The tarot is considerably more complex to consult than is the *I Ching*. Tarot philosophy is also considerably more complex. On the other hand, those who use it maintain that its answers can be more specific and detailed than those given by the *I Ching*. As with the *I Ching*, an essential prerequisite seems to be that one approach the tarot with respect. Unlike the *I Ching*, it is recommended that one *not* perform tarot readings for oneself, for there is danger of losing objectivity in doing so.

Both the *I Ching* and the tarot have given very impressive results for the authors, though our experience with the latter is limited. "The Oracle," as the *I Ching* is often called, has more than once been particularly helpful when we were confronted with the necessity of making difficult decisions. We have a great sense of reverence for these procedures, however, and so use them only sparingly. As Jung himself wrote, late in his life, "You know I have not used it [the *I Ching*] in more than two years now, feeling that one must learn to walk in the dark, or try to discover (as when one is learning to swim) whether the water will carry one."[204]

APPENDIX II

SYNCHRONICITY

AND PROBABILITY

Perhaps we need to be much more radical in the explanatory hypotheses considered than we have allowed ourselves to be heretofore. Possibly the world of external facts is much more fertile and plastic than we have ventured to suppose. . . .

E. A. BURTT

*The Metaphysical Foundations
of Modern Science*

The probabilistic interpretation of synchronicity, that all so-called synchronicity is due to the vicissitudes of mere chance, is not dealt with in depth in the main text. One reason for this is that we wish to avoid the tiresome academic writing style requiring that one answer one's opponents' objections on every page. Beyond this, we believe along with Michael Shallis that arguments based on probability are not sound.[205] Let us briefly explain.

To know the mathematical probability of a specific event requires a knowledge of all possible events that can occur in the situation in question. For instance, if four horses run a race we can say that, given no other information about the horses, the chances of a particular one winning are 1 in 4. If we know that one horse has never lost a race while the other three have never won, we can increase our bet on the former horse, but we cannot establish a precise mathematical probability for its winning. The variables that can influence real horses are too complex to allow mathematical precision. Suppose, to go further, that we do not know how many horses are racing or that all entries are, in fact, horses or worse yet that the race will even occur. Now what are our probabilities? Without precise prior knowledge of the situation we simply cannot calculate probabilities for it.

How many types of insects might in theory have flown against the window of Jung's consulting room during his client's recitation of her dream? What were the chances that she would have such a dream? How likely was it that she would have told it to Jung at that moment? How often did Jung notice insects tapping against his window? It simply is not possible to calculate even a rough probability for such an event. And, of course, any such calculation must fail to include the subjective meaning of the event to the client that gave it its numinosity.

Let us examine an instance reported by Arthur Koestler in *The Challenge of Chance*.[206] This case concerned a publisher, Jeffery Simmons, of W. H. Allen & Company. After twenty-five years in publishing, Simmons found himself for the first time in the position of having to destroy his entire stock of a book by pulping it—selling the books for their value as mere paper pulp. He did not know where to find a paper mill to buy the books, so he went upstairs to ask the production manager. Though the manager did not know either, a young boy from the warehouse happened by at that moment and, hearing the conversation, volunteered the name of the Phillips Mills, with which he was familiar because it was located near his home. Using the production manager's phone, Simmons asked the receptionist to obtain that firm's phone number. She answered, "Their representative is here." Simmons at first thought she was joking, but the representative had, indeed, walked in just a few seconds before. He was an old man whose business rounds took him by the publishing company almost daily, though he had not previously stopped there. He had entered that day on an impulse and, after conducting the day's business, did not return again.

The final chapter of this story involved Koestler himself. Simmons had read of Koestler's interest in obtaining accounts of unusual co-incidences and was debating with himself whether to write him about this one when he was visited by a client, Viscount Maugham, the nephew of Somerset Maugham, on a business matter. What made up Simmons's mind was that Maugham had just come from Koestler's home.

According to the most common theory of probability, the frequency theory, the likelihood of any event is based upon how often it has occurred in the past. But Mr. Simmons had never before pulped a book, nor had the representative from Phillips Mills ever visited his publishing company, so we are at an immediate loss about how to compute probabilities for these events. This says nothing of the like-lihood that the boy from the warehouse would happen by with the necessary information or, more importantly, that *all these events would occur at essentially the same time*, for otherwise nothing un-usual would have transpired. All this is compounded by the con-currence of the arrival of Viscount Maugham at the Simmons residence, directly from Arthur Koestler's home, as Simmons was debating whether to send an account of these events to Koestler himself.

In fact, statistical probabilities may be calculated in the real world in only a very few types of situations. One is in the laboratory, where the scientist or engineer has a high degree of control over all con-ditions. Another is in relating mass statistics concerning large num-bers of similar events, such as how many Big Mac's will be sold in the United States on a particular Fourth of July. Exactly who will buy the burgers is not known. Still another situation in which prob-abilities can be highly specified involves quantum statistics, which deal with subatomic events. None of these situations applies to most synchronistic coincidences, which are never under control and are always unique to the individual.

A more fundamental problem is the nature of probability itself. We all have an intuitive notion of it, but to use an example from Bertrand Russell, what does it mean to say that the probability of a plane crash on a particular flight, say, to Paris, is 1 in 10,000?[207] According to frequency theory, it might mean that the plane has only crashed on the average of once in every ten thousand flights! But this cannot be correct. If the plane had previously crashed, we would not be flying on it. In any event, we are talking about the future—the next flight—not the past. Each flight is unique, with its own atmospheric conditions, pilot and copilot, and so on. How can the past apply to this flight? The same objections apply if the 1 in 10,000 probability is based on all flights of all planes of this type. How can we be sure that such a variety of flights under diverse real conditions has anything whatever to do with this particular flight that we are perhaps about to get on?

Frequency theory runs into serious trouble if, as is usually the case with synchronicity, the event in question never happened before, and there is no sense in talking about its frequency in the first place. The queen of Spain once dreamed that there was a pig in the throne room. Much to her surprise, upon descending the stairs in the morning she found a pig! We are told that no pig had ever been seen in the throne room before, so how can we state the probability of this event, to say nothing of the probability of the dream itself?

Let us return to our plane and the probability of 1 in 10,000 that it will crash. In fact, we know that the plane either will crash or it will not crash. If it crashes, is it not true that prior to the flight the probability was 1 in 1 (100 percent) that it would crash? And if it does not, is it not true that the probability was 0? Given that the event will definitely happen or not happen, how can any probabilities other than 1 in 1 or 0 be correct?

In his book, *Theories of Probability: An Examination of Foundations*, mathematician T. R. Fine reviews no fewer than six distinct theories of the nature of probability. These include the frequency interpretation, limited relative-frequency interpretation, algorithmic theory, and so on. Fine considers whether there is sufficient justification to apply any of them legitimately to the real world. He concludes that there is not.[208]

Aside from all this, it is our view that no amount of arguing will finally settle the question of whether all coincidences amount to nothing more than mere happenstance. The real issue is how one chooses, or has learned, to interpret the experience of a sometimes indefinite reality. There is no final court of appeal to judge what is true and what is not true. If there were, philosophy would have been at an end thousands of years ago.

It has been the strong prejudice of the scientific intelligentsia for the past three hundred years to favor reductionistic interpretations. The situation is changing, but much of its inertia is still with us. The point here, however, is that the issue of probability is one of choice and not of evidence; there is no hard evidence in the matter of synchronicity, and arguments can be built in both directions. Most of us are all too familiar with the reductionistic frame of mind. To appreciate the other point of view, the one represented in this book, one need not believe it, but at a minimum one must, to use John Lilly's words, "simulate the belief" and see where it leads.

APPENDIX III
THE PSI-FIELD HYPOTHESIS

The main ideas set forth in the psi-field hypothesis seek to advance the simplest possible scheme of thought capable of uniting the prima facie disparate domains of physical and of living nature. Such a scheme calls for an interactive evolutionary dynamic that is neither fully deterministic, nor punctuated by fully random events. The required dynamic is probabilistic but oriented; its probabilities constrained by order-generating interconnections among the evolving systems. Such interconnections are best conceptualized in terms of continuous fields, and the required properties of the fields are best satisfied by the holographic mode of information storage and transmission. The physical roots of the indicated holofield, in turn, are most reasonably sought in the Dirac sea of the quantum vacuum. These ideas are presented in detail in my recent books, *The Creative Cosmos* and *The Interconnected Universe*.

A scheme that would exhibit the quantum vacuum as a universal interconnecting field must show that it is an energy field with a substructure. This is consistent with evidence already in hand. The relevant finding is the high degree of complexity that under-lies the interactions of quanta. This suggests either that quanta themselves are compound entities, with an internal structure that accounts for the specific complexity of their interactions, or that the structure of the field in which they are embedded has the re-quired degree of complexity. Both these assumptions have been ex-plored by theoretical physicists, and it has appeared that there is no reliable evidence for assuming that quanta would be complex en-tities in themselves. It is known, on the other hand, that the quan-tum vacuum is a cosmically extended electromagnetic field. In this field the interaction of massless charges creates 'matter' in the form of mass, which in turn creates gravitation and produces the attract-ive and repulsive forces associated with nuclear fields. This field does have a substructure, though mainstream physics views it as homogeneous and isotropic, filled with purely random fluctuations (Zitterbewegungen).

The assumption that the massless charges that constitute the fundamental units of the observable universe would be embedded in a vacuum field with some type of substructure is intrinsically reasonable. The question is mainly whether that substructure inter-acts in significant ways with the phenomena that appear in it. In stochastic electrodynamics (SED) the assumption that it does inter-act produces remarkably accurate accounts of quantum phenom-ena, without requiring the complex auxiliary assumptions of contemporary quantum mechanics.

In view of considerations such as these, we choose the second option: quanta, in this view, are embedded in a complex field. The next question concerns the information-richness of the quantum vacuum, more exactly, of its zero-point electromagnetic field (ZPF). As just noted, physicists view the ZPF as homogeneous, iso-tropic, and Lorentz-invariant. This is to satisfy the finding that con-stant motion through the ZPF does not give rise to asymmetries (though accelerated motion produces the distortions predicted by Davies and Unruh and shown to be the physical basis of inertia by Puthoff, Rueda, and Haisch). The query that needs to be posed is whether asymmetries of motion detected by physical instruments are the only conceivable kinds of effects that the field would produce on observable phenomena. The concept of the ether was discarded for failing to produce effects of this kind, yet as Michelson himself observed, this should not have done away with the concept of a continuous underlying field that would function as a charge-conveying and information-transmitting medium.

The central idea here is that of an information-rich subquantum field as the most reasonable heuristic device required to bring seemingly disparate physical and biological phenomena within the compass of an integrated theoretical framework.

We now consider the processes whereby a substructure is created within the ZPF. As is well known, classical electrodynamics predicts that a fluctuating electric charge emits an electromagnetic radiation field. In their interaction with the ZPF, electrically charged quanta are believed to produce secondary electromagnetic fields, and these fields must be universally extended. The energy associated with the fields gives rise to fluctuations in charged particles that propagate relativistically, at or near the speed of light. Zitterbewegung is assumed to be random, satisfying the tenet of homogeneity, isotropy and Lorentz-invariance for the ZPF. However, the present hypothesis maintains that the fields created by the motion of quanta within the vacuum have a non-isotropic, non-homogeneous and non-Lorentz-invariant component. This component consists not of the familiar transverse electromagnetic waves, but of longitudinally propagating scalar ('Tesla') waves. In this regard the motion of charged particles through the ZPF approximates the action of a monopole antenna: it alternately charges and dis-charges local regions of the primary electromagnetic field. The thus triggered longitudinal waves alternately compress and rarefy the virtual-particle gas of the vacuum. In consequence, the structure of the ZPF becomes mediated by secondary scalar fields, generated by quanta. We envisage the ZPF as a structured field with a scalar-mediated electromagnetic spectrum.

In regard to its scalar component, the ZPF is a continuum of which each point is defined by a corresponding magnitude. At each point of the field the non-random flux is a local scalar wave within the massless charge spectrum. The magnitude at a given point is an n–dimensional virtual-state flux. The field is a continuum of stresses and potentials. Its stress energies can be expressed in terms of geometrodynamics as an electrostatic scalar potential.

We now consider the specifics of the scalar field-component of the vacuum substructure. The first thing to note is that scalar waves are not of the kind that satisfy D'Alembert's equation, that is, they are not similar to light and sound waves. This is because the characteristic feature of a D'Alembert equation is the occurrence of a second time-derivative term of the wave amplitude, and generally such a term is a consequence of the inertial properties of matter. In a medium such as the quantum vacuum, these properties would not apply; vacuum waves are better represented by fundamental equations that contain only first-order time-derivative terms. But the only kinds of first-order time-derivative equations governing linear wave propagations are Schrödinger wave equations. Thus we hypothesize that the scalars mediating the ZPF approximate Schrödinger waves.

In regard to at least four factors, this additional postulate satisfies the memory-functions we require to be associated with an information-rich ZPF. First, Schrödinger waves are linear, thus allowing interfering wave-trains to superpose, conserving rather than destroying phase information. Second, unlike holographic wave patterns composed of D'Alembert waves that can only be recorded on plates, interference patterns created by Schrödinger type waves require the entire medium in which the waves propagate. This allows for vastly more information storage by Schrödinger wave holograms than by holograms based on D'Alembert waves. Third, the diffraction of Schrödinger waves through a fixed hologram can create time-varying information, whereas only time-invariant information can be recorded with D'Alembert type interference patterns. And fourth, items of time-varying information injected into the holographic medium through a system of independent point-like sources can be recovered in a Schrödinger type hologram in the proximity of the sources, whereas information recovery in D'Alembert-type holograms is possible only by focalization devices for the diffracted waves.

We next consider the velocity of Schrödinger type secondary scalar wave propagations in the vacuum. Schrödinger waves generated by periodic emissions are known to propagate with velocities proportional to the square root of their specific frequencies. In the vacuum these velocities cannot be held limited by the constant that applies to the propagation of charged masses in the electromagnetic spectrum. In relativity theory the light-constant functions as an axiom of invariance: in a normal coordinate system the speed of light, c, is to be chosen so as to remain constant regardless of the curvature of spacetime; and the unit of time is to be chosen so that the speed of light within the local system of coordinates becomes equal to one. In the present hypothesis we obtain a physical basis for the observed value of c: this is the finite electromagnetic permittivity/permeability of the vacuum. If the value of c is inversely proportional to the square-root of the product of the vacuum's electrostatic permittivity and magnetic permeability, assuming that the electric and magnetic components refer to comparable vacuum constraints gives a physical basis for the limited velocity of charged masses (quanta) but leaves massless charge propagations (vacuum scalars) unaffected.

Scalars do not propagate differently from classical electromagnetic waves. In a seminal paper published in 1903, E. T. Whittaker has shown that longitudinal waves propagate with a finite velocity that may be enormously greater than the speed of light. Since scalars are longitudinal waves, their propagation is proportional to the mass-density of the medium in which they propagate. Mass-density defines the local electrostatic scalar potential of the vacuum: it is a variable quantity, higher in regions of dense mass, in or near stars and planets, and lower in deep space (a variation due to the increase in vacuum flux intensity by the accumulation of charged masses). Consequently scalars travel faster through matter-dense regions of the vacuum than in deep space, much as longitudinally propagating sound waves travel faster in a dense medium such as water than in a thin medium such as air (and as Whittaker's longitudinal gravitational waves travel faster than light waves).

The scalar-mediated ZPF can produce an effect on observable phenomena due to the interaction of the massless scalars with charged masses (quanta). The interaction of scalars and quanta approximates a two-way Fourier transformation: the vacuum encodes the coefficients of the interfering scalar wavefronts produced by the motion of quanta. In so doing the vacuum carries out the equivalent of the forward Fourier-transform: it translates a pattern from the spatiotemporal to the spectral domain. In the inverse transform — from the spectral to the spatiotemporal domain — the interference patterns encoded in the vacuum 'inform' the motion of quanta in space and time. With some simplification we can say that there is an ongoing read-in by quanta of their spatio-temporal motion into the vacuum, and a similarly ongoing read-out of the corresponding information.

The two-way translation between quanta and the quantum vacuum is universal in scope but specific in occurrence: quanta retranslate from the vacuum only those transforms that match their own quantum states. Such specificity is due to the fact that in Fourier transformations the reverse transforms are the exact inverse of the forward transforms. Retranslation selectivity ensures that quanta are not overwhelmed by the information conveyed through the scalar-mediated ZPF. Quantal states are affected only by the spectral transform of their own wave functions. In a simplified terminology we can say that each quantum 'reads out' from the vacuum only the information that it has itself 'read in'.

My recent book, *The Interconnected Universe* applies the above postulates to both microscale systems, of the order of Planck's constant; and macroscale systems, consisting of large ensembles of microscale components. The thesis is that the motion of quanta generates scalar patterns in the ZPF, and the thereby modified topology of the field interacts with the motion of quanta in the corresponding quantum state. The isolation of a quantum becomes an abstraction, even if in individual cases a valid one. Fundamentally, quanta are non-isolable, constantly interacting entities, in which constraints on degrees of freedom, introduced by interactions with the scalar-mediated ZPF, appear on the level of ensembles. This tenet safeguards the formalisms of quantum mechanics in regard to individual systems, but qualifies them with interaction-generated probabilistic irreversibilities for populations of systems.

Beyond the Planck-domain, it is known that quantum or subquantum- level fluctuations in macroscale systems are damped by large- scale regularities that obey dynamical laws. Interaction effects could appear nevertheless, as Poincaré resonances manifested in populations of the component quanta are amplified by the chaos- dynamics of the system. The state of deterministic chaos is highly initial-condition dependent. Unmeasurably small changes in initial conditions—or, in persistent systems, in parametric conditions—are registered as strange attractors amplify the changes into dynamic inputs to the evolutionary trajectory. Our hypothesis suggests that in a chaotic state macro-scale Poincaré systems are sufficiently interactive to produce measurable interaction-effects. These interactions involve not the quantum state of the individual component but the 3n–dimensional configuration-space of the whole multiquantal system. As a result, we can expect that in dynamically indeterminate ('chaotic') states, macroscale systems are 'in-formed' with the ZPF-conveyed Fourier-transform of their 3n–dimensional configuration space.

The above tenet applies with particular force to living systems. These systems persist far from thermodynamic equilibrium, in states that are inherently both unstable and ultrasensitive. Organic sensitivity extends to energies and radiations well below the thermal threshold of chemical reactions, expressed as kT, a function of the Boltzmann constant and absolute temperature. Experimental findings indicate that the condition of the organism is not dependent uniquely on the equilibrium thermodynamics of thermal energy exchanges and tissue heating but involves quantum states and resonant responses.

The basic idea, in summary, is that of quantum/vacuum interaction (QVI) as the 'simplest possible scheme' capable of unifying the observed facts in the physical and in the life sciences.

REFERENCES

1. Laszlo, E. (1993). *The Creative Cosmos*. Edinburgh: Floris Books.
2. Laszlo, E. (1995). *The Interconnected Universe*. New York, London, and Singapore: World Scientific.

CASES OF SYNCHRONISTIC COINCIDENCES

REFERENCES

PREFACE

1. Broad, L. (1958) *Winston Churchill*. New York: Hawthorn.
2. Toland, J. (1976) *Adolf Hitler*. New York: Doubleday, 60.
3. Jung, C. G. (1973) *Synchronicity: An acausal connecting principle*. CW, vol. 8. Princeton, N.J.: Princeton Univ. Press.
4. Ibid., 22.
5. Ibid.

INTRODUCTION

6. Precope, J., trans. (1952) *Hippocrates on diet and hygiene*. London, 174.
7. Thompson, W. I. (1981) *The time falling bodies take to light*. New York: St. Martin's Press.
8. Barbour, I. G. (1974) *Myths, models, and paradigms*. New York: Harper & Row.
9. Whitehead, A. N. (1967) *Science and the modern world*. New York: Free Press, 24.
10. Novalis. (1798) *Pollen and Fragments*. Cited in R. Bly (1980) *News of the universe*. San Francisco: Sierra Club Books, 48.
11. Haywood, J. W. (1984) *Perceiving ordinary magic*. Boulder and London: Shambhala, 59.
12. Comfort, A. (1984) *Reality and empathy*. Albany, N.Y.: State Univ. of New York Press.
13. Barnes, E. W. (1933) *Scientific theory and religion*. Cambridge: Cambridge Univ. Press, 583.
14. Descartes, R. (1951) *Meditations on first philosophy*, trans. L. J. Lafleur. New York: Bobbs-Merrill. Original work published in 1641.

15. Bohm, D. (1980) *Wholeness and the implicate order*. London: Routledge & Kegan Paul.
16. Comfort, *Reality and empathy*.
17. Ibid., 25.
18. Jantsch, E. (1980) *The self-organizing universe*. New York: Pergamon.
19. Haywood, *Perceiving ordinary magic*, 174.
20. Hardy, A., Harvie, R., and Koestler, A. (1973) *The challenge of chance*. New York: Random House, 194–95.
21. Gamow, G. (1966) *Thirty years that shook physics*. Garden City: Anchor, 64.
22. Thompson, *The time falling bodies take to light*.
23. Campbell, J. (1986) *The inner reaches of outer space: Metaphor as myth and religion*. New York: Alfred van der Marck, 17.
24. Ibid., 18.
25. Smuts, J. (1926) *Holism and evolution*. New York: Macmillan.
26. Watson, L. (1980) *Lifetide: The biology of consciousness*. New York: Simon & Schuster.
27. Bohm, D. (1982) Nature as creativity: A conversation with David Bohm. *ReVision* 5. (February): 35–40, 39.

CHAPTER 1

28. Kammerer, P. (1919) *Das gesetz der serie*. Stuttgart and Berlin: Versags-Anstalt.
29. Koestler, A. (1972) *The case of the midwife toad*. New York: Random House, 136.
30. Kammerer, *Das gesetz der serie*.
31. Koestler, A. (1972) *The roots of coincidence*. New York: Random House, 87.
32. Koestler, A. *The case of the midwife toad*, 139–140.
33. Briggs, J. P., and Peat, F. D. (1989) *Turbulent mirror: An illustrated guide to chaos theory and the science of wholeness*. New York: Harper & Row.
34. Shallis, M. (1983) *On time*. New York: Schocken Books.
35. Koestler, *The case of the midwife toad*.

CHAPTER 2

36. Comfort, *Reality and empathy*.
37. Bohm, *Wholeness and the implicate order*, 189.
38. d'Espagnat, B. (1979) Quantum theory and reality. *Scientific American* (November): 158–81.
39. Ibid.
40. Herbert, N. (1985) *Quantum reality*. New York: Anchor/Doubleday.
41. Bohm, *Wholeness and the implicate order*.
42. Comfort, *Reality and empathy*.
43. Bohm, *Wholeness and the implicate order*.
44. Shabistari, M. (1974) *The secret garden*. Trans. Johnson Pasha. New York: E. P. Dutton.
45. Hardy, Harvie, and Koestler, *The challenge of chance*.

REFERENCES

46. Ibid., 173.
47. Sheldrake, R. (1981) A *new science of life*. London: Blond & Briggs.
48. Sheldrake, R. (1987) Part I: Mind, memory and archetypes—Morphic resonance and the collective unconscious. *Psychological Perspectives* 18 (January): 9–25, 12.
49. Sheldrake, *A new science of life*.
50. Sheldrake, R. (1988) *The Presence of the past*. New York: Random House.
51. Ibid.
52. Ibid.
53. Ibid.
54. Sheldrake, R. and Bohm, D. (1982) Morphogenetic fields and the implicate order: A conversation. *ReVision* 5 (February): 41–48.
55. Laszlo, E. (1987) The psi-field hypothesis. *IS Journal* 4: 13–28.
56. Bohm, D., and Peat, F. D. (1987) *Science, order, and creativity*. New York: Bantam.
57. McWaters, B. (1981) *Conscious evolution: Personal and planetary transformation*. San Francisco: Institute for the Study of Conscious Evolution, 64.
58. Jung, *Synchronicity*.
59. Grof, S. (1987) *The adventure of self-discovery*. Albany, N.Y.: State Univ. of New York Press, 152.
60. Von Franz, M-L. (1980) *Projection and recollection in Jungian psychology*. Lasalle, Ill.: Open Court.
61. Padfield, S. (1981) Archetypes, synchronicity, and manifestation. *Psychoenergetics* 4 (January): 77–81.
62. Padfield, S. (1980) Mind-matter interaction in the psychokinetic experience. In D. D. Josephson and V. S. Ramachandran, eds., *Consciousness and the physical world*. New York: Pergamon Press, 167.
63. Padfield, Archetypes, synchronicity, and manifestation, 80.
64. Hardy, Harvie, and Koestler, *The challenge of chance*.
65. Koestler, *The roots of coincidence*.
66. Padfield, Mind-matter interaction in the psychokinetic experience, 169.
67. Bailey, A. (1972) *The yoga sutras of Patanjali*. Lucis, 394.
68. Padfield, Archetypes, synchronicity, and manifestation, 80.
69. Stapledon, O. (1972) *Starmaker*. Baltimore: Penguin, 79. Original work published 1937.
70. Chester, A. N. (1981) A physical theory of psi based on similarity. *Psychoenergetics* 4 (February): 89–111.
71. Teilhard de Chardin, P. (1959) *The phenomenon of man*. New York: Harper & Row.
72. Ibid., 65.
73. Ibid., 65.
74. Ibid., 60.
75. Progoff, I. (1973) *Jung, synchronicity, and human destiny: Acausal dimensions of human experience*. New York: Dell.
76. Jantsch, *The self-organizing universe*.
77. Koestler, A. (1978) *Janus: A summing up*. New York: Random House.
78. Ibid.

79. Ibid., 270.
80. Loye, D. (1983) *The sphinx and the rainbow*. Boulder: Shambhala.
81. Targ, R., and Harary, K. (1984) *The mind race*. New York: Villard Books.
82. Goldberg, P. (1983) *The intuitive edge*. Los Angeles: J. P. Tarcher.
83. Chu Hsi. (1922) *The philosophy of human nature by Chu Hsi*. Trans. J. Percy Bruce. London: Probstrain.

CHAPTER 3

84. Dubois, R. L. (1960) *Pasture and modern science*. London: Heinemann.
85. Jackson, J. H. (1864) Some clinical remarks on cases of defects of Expression (by words, writing, signs, etc.) in diseases of the nervous system. *Lancet*, 2: 604.
86. Bruner, J. (1962) *On knowing: Essays for the left hand*. Cambridge, Mass.: Harvard Univ. Press, 2.
87. Loye, *The sphinx and the rainbow*.
88. Honegger, B. (1979) *Spontaneous waking-state psi as interhemispheric verbal communication*. San Francisco: Washington Research Center.
89. Jaynes, J. (1976) *The origin of consciousness in the breakdown of the bicameral mind*. Boston: Houghton Mifflin.
90. Penfield, W., and Perot, P. (1963) The brain's record of auditory and visual experience: A final summary and discussion. *Brain* 86: 595–702.
91. Honegger, *Spontaneous waking-state psi as interhemispheric verbal communication*.
92. Ibid., A12.
93. Jung, C. G. (1962) Commentary on the secret of the golden flower, in Wilhelm, R., *The secret of the golden flower*. Trans. C. F. Baynes. New York: Harcourt, Brace & World.
94. Russell, P. (1983) *The global brain*. Los Angeles: J. P. Tarcher, 214.
95. Ibid., 214.
96. Orme-Johnson, D. W. (1977) Higher states of consciousness: EEG coherence, creativity and experiences of the siddhis. *Electroencephalography and Clinical Neurophysiology* 4: 581.
97. Pribram, K. (1984) The holographic hypothesis of brain function: A meeting of minds, in S. Grof, *Ancient wisdom and modern science*. Albany, N.Y.: State Univ. of New York Press.
98. Abu-Mostafa, Y., and Psaltis, D. (1987) Optical neural computers. *Scientific American* 256 (March): 88–95.
99. Loye, *The sphinx and the rainbow*.

CHAPTER 4

100. Jung, *Synchronicity*, 34.
101. Jung, *Synchronicity*, 22.
102. Jung, *Synchronicity*, 10.
103. Schopenhauer, A. (1851) *Sämtliche Werke*, vol. 8.

ᅟ

104. Schopenhauer, *Sämtliche Werke*, 225.
105. Shabistri, *The secret garden*.
106. Jung, *Synchronicity*.
107. Thompson, *The time falling bodies take to light*.
108. Campbell, J. (1959) *The masks of god: vol. 1, primitive mythology*. New York: Viking, 263.
109. Meade, M. (1988) "I'll answer in three days": An interview with Michael Meade. *Inroads: A Journal of the Male Soul* 1 (January): 4–13, 7.
110. Progoff, I. (1973) Jung, synchronicity, and human destiny. New York: Dell, 83.
111. Homer. (1950) *The Iliad*, trans. E. V. Rieu. Baltimore: Penguin, 446–47.
112. Vaughan, A. (1979) *Incredible coincidence*. New York: J. B. Lippincott.

CHAPTER 5

113. Otto, W. F. (1954) *The homeric gods*. Trans., M. Hadas. New York: Pantheon.
114. von Franz, *Projection and recollection in Jungian psychology*.
115. Ibid., 61.
116. Kuhn, T. H. (1962) *The structure of scientific revolutions*. Chicago: University of Chicago Press.
117. Kerenyi, K. (1976) *Hermes guide of souls: The mythologem of the masculine source of life*. Dallas: Spring, 104.
118. Ibid., 13.
119. Ibid., 84.
120. Rama, Swami S. S., Ballentine, R., and Ajaya, Swami. (1976) *Yoga and psychotherapy: The evolution of consciousness*. Honesdale, Pa.: Himalayan Press.
121. Wilber, K. (1983) *Eye to eye: The quest for the new paradigm*. New York: Doubleday, 122.
122. Stein, M. (1983) *In midlife: A Jungian perspective*. Dallas: Spring.
123. Jung, *Synchronicity*.
124. Teit, A., Farrand, L., Gould, M. K., and Spinden, H. J. (1917) *Folk-tales of Salishan and Sahaptin tribes*. F. Boas, ed.; Lancaster, Pa.: G. E. Stechert.
125. Graves, R. (1955) *The Greek myths*. Baltimore: Penguin, 63.
126. Burkert, W. (1985) *Greek religion*, trans. John Raffan. Cambridge, Mass.: Harvard Univ. Press.
127. Otto, *The homeric gods*, 123.
128. Hillman, J. (1979) *The dream and the underworld*. New York: Harper & Row.
129. Kerenyi, *Hermes guide of souls*.
130. Campbell, *The masks of god*.
131. Kerenyi, *Hermes guide of souls*, 51.
132. Burkert, *Greek religion*, 158.
133. Campbell, *The masks of god*, 416.
134. Kerenyi, *Hermes guide of souls*.
135. Pelton, R. (1980) *The trickster in West Africa: A study in mythic irony and sacred delight*. Berkeley: Univ. of California Press.

136. Campbell, *The inner reaches of outer space*, 55.
137. Globus, G. (1986) Three holonomic approaches to the brain. In David Bohm: *Physics and beyond*, B. Hiley and D. Peat, eds. London: Routledge & Kegan Paul.
138. Otto, *The homeric gods*, 121.
139. Jung, C. G. (1967) *Alchemical studies*. CW, vol. 13. Princeton, N.J.: Princeton Univ. Press, 235.
140. Brown, N. O. (1969) *Hermes the thief*. New York: Vintage Books.
141. Von Franz, M-L. (1979) *Alchemical active imagination*. Dallas: Spring, 94.
142. Pelton, *The trickster in West Africa*, 49.
143. Ibid., 141.
144. Aurobindo, S. (1974) *Savitri*. Pondicherry, India: Sri Aurobindo Ashram Press, 20.
145. Pelton, *The trickster in West Africa*, 60.

CHAPTER 6

146. Otto, *The homeric gods*, 117.
147. Pelton, *The trickster in West Africa*.
148. Whitmont, E. C. (1969) *The symbolic quest: Basic concepts of analytical psychology*. New York: Harper & Row.
149. Jung, C. G. (1959) *Four archetypes: Mother, rebirth, spirit, trickster*. Bollingen Series, vol. 9, pt. 1. Princeton, N.J.: Princeton Univ. Press, 147.
150. Peat, D. (1987) *Synchronicity: The bridge between matter and mind*. New York: Bantam, 18.
151. Brown, *Hermes the thief*.
152. Kerenyi, *Hermes guide of souls*, 84.
153. Zimmer, H. (1971) *The king and the corpse*. Bollingen Series XI. Princeton, N.J.: Princeton Univ. Press.
154. Hoffman, D. (1961) *Form and fable in American fiction*. London: Oxford University Press.
155. Twain, M. (1884/1977) *The adventures of Huckleberry Finn*. Sculley Bradley et al., eds. New York: Norton, 72.
156. Otto, *The homeric gods*.
157. Jung, C. G. (1959) *Aion*. CW, vol. 9, pt. II. Princeton, N.J.: Princeton Univ. Press.
158. Jung, *Synchronicity*, 23.
159. Erdoes, R., and Fire, J. (1971) *Lame Deer: Seeker of visions*. New York: Simon & Schuster.
160. Ibid., 225.
161. Stein, *In midlife*.
162. Pelton, *The trickster in West Africa*, 60.
163. Johnson, R. (1986) *Inner work*. San Francisco: Harper & Row, 59.
164. Progoff, I. (1980) *The practice of process meditation*. New York: Dialogue House.
165. Ibid.

166. Campbell, J., and Moyers, B. (1988) *The power of myth*. New York: Parabala Magazine, Mystic Fire Video.
167. Vissell, B., and Vissell, J. (1984) *The shared heart*. Aptos, Calif.: Ramira .
168. Progoff, *Jung, synchronicity, and human destiny*.
169. Hardy, Harvie, & Kosetler, *The challenge of chance*.
170. Cox, W. E. (1956) *Precognition: An analysis II*. ASPR 50: 99–109.
171. Weaver, W. (1963) *Lady luck: The theory of probability*. New York: Anchor.
172. Jung, C. G. (1965) *Mysterium coniunctionis*. CW, vol. 14, pt. II. Princeton, N.J.: Princeton Univ. Press, 535.
173. Serrano, M. (1968) *Jung and Hess: A record of two friendships*. New York: Schocken, 86.
174. Von Franz, *Alchemical active imagination*.
175. Berman, M. (1981) *The reenchantment of the world*. Ithaca, N.Y.: Cornell Univ. Press.
176. Von Franz, *Alchemical active imagination*, 83.
177. Rama, Swami S.S. (1987) Developing strength and willpower. *Dawn* 6 (March): 6–14.
178. The Mother. (1972) *Questions and answers: 1950–1951*. Pondicherry, India; Sri Aurobindo Ashram Press, 233.
179. Tirtha, Swami R. (1978) *The practical Vedanta*. Honesdale, Pa.: Himalayan Press, 36–37.
180. Ibid., 6–7.
181. Whitmont, E. C. (1982) *The return of the goddess*. New York: Crossland, 231.
182. Ibid., 231.
183. Lilly, J., and Lilly, A. (1976) *The dyadic cyclone*. New York: Simon & Schuster, 27.
184. Whitmont, *The return of the goddess*.
185. Lilly and Lilly, *The dyadic cyclone*, 25–26.
186. Spangler, D. (1975) *The laws of manifestation*. Marina Del Ray, Calif.: DeVross, 55.
187. Leggett, T. (1960) *A first Zen reader*. Rutland, Vermont: Tuttle, 178.
188. Von Franz, M-L. (1980) *Alchemy: an introduction to the symbolism and the psychology*. Toronto: Inner City, 236.
189. Jung, Commentary on the secret of the golden flower, 124.
190. Von Franz, *Alchemy*, 237.
191. Otto, *The homeric gods*, 115.
192. Kerenyi, *Hermes guide of souls*, 91.
193. Lucian. (2nd century A.D.) *Dialogues of the gods*.
194. (1963) *New Testament Apocrypha II*. (English Translation) London, 88.

APPENDIX I

195. See, for example, Wilber, K. (1981) *Up from Eden*. Garden City, New York: Doubleday.
196. Harner, M. (1980) *The way of the shaman*. New York: Harper & Row, 89.

197. Seligman, K. (1948) *Magic, supernaturalism and religion.* New York: Random House.
198. Chou, H. (1979) Chinese oracle bones. *Scientific American* 240 (April): 134–44, 148–49.
199. Loewe, M., and Blaker, C. (1981) *Oracles and divination.* Boulder, Colorado: Shambhala, 152.
200. Ibid., 152–53.
201. Loewe and Blaker, *Oracles and divination,* 7.
202. Wilhelm, R. (1950) *The I Ching or book of changes.* Princeton, N.J.: Princeton Univ. Press.
203. Jung, C. G. Commentary on "The Secret of the Golden Flower," 141.
204. Jung, C. G. (1945) Personal letter to Dr. Liliane Fry-Rohn written on February 27. Cited in A. Jaffé, *C. G. Jung: Word and symbol.* Princeton, N.J.: Princeton Univ. Press, 45.

APPENDIX II

205. Shallis, *On time.*
206. Hardy, Harvie, and Koestler, *The challenge of chance.*
207. Russell, B. (1961) *Human knowledge.* London: Allen & Unwin.
208. Fine, T. R. (1973) *Theories of probability: An examination of foundations.* New York: Academic Press.

INDEX